PLANETARY GEOLOGY

PLANETARY GEOLOGY

**John Guest with Paul Butterworth,
John Murray and William O'Donnell**

David & Charles
Newton Abbot London

To Miss Peabody

British Library Cataloguing in Publication Data

Planetary geology.
 1. Solar system   2. Geology
 I. Guest, John Edward
 559.9   QB505

ISBN 0-7153-7739-6

80 007204

*DF
.D
P*

Typeset by Trade Linotype Ltd, Birmingham
and printed in Great Britain
by Biddles Limited, Guildford
for David & Charles (Publishers) Limited
Brunel House Newton Abbot Devon

Contents

The Moons of Mars — The Moons' Motion, Cratering and Albedo — The Grooved Moon

Introduction

Having satisfied his urge to explore the continents of Earth it is not surprising that Man has turned his attention to the planets. Suddenly, many of our planetary neighbours have ceased to be distant objects viewed through a telescope and have become worlds that we can see at close quarters. In the first phase of exploration of the Moon and the terrestrial planets Mars and Mercury, geologists have played an important part: this is particularly true for the Moon, where careful investigation of potential landing sites was an integral part of the NASA plan to land men on the Moon; geological, including geophysical and geochemical investigations, have been, and still are, a major outcome of exploration.

Within the last fifteen years the areas studied by the Earth scientist have expanded to encompass not just one member of the Solar System, the Earth, but all the planets so far explored; and during the next decade or so we may reasonably expect to have explored to a greater or lesser extent most of the planets and satellites in the Solar System.

Geological examination of the planets tends to follow a systematic pattern. Our first views of the planet under investigation are of relatively low resolution; with time and successive new missions our view becomes progressively more detailed. On the Moon initial studies were carried out using telescopic pictures and were followed by observations made by spacecraft in orbit around the Moon, unmanned soft-landers and finally manned landings on the surface where field observations were made and rocks collected for return to Earth. Studies of Mars have followed much the same pattern, the latest mission, that of Viking, having produced large amounts of data from orbit as well as from surface studies made by two soft-landers. Exploration of Mercury is still at an early stage and relies entirely on the Mariner 10 mission which flew by Mercury providing photographic data comparable to that available from telescopic observation of the Moon and covering just under 50% of Mercury's surface. The next stage in exploration of the Moon will almost certainly be to map the chemistry of the surface and make other geophysical measurements from an orbiter placed in low altitude around our satellite; a similar mission may also be contemplated for Mercury, whereas for Mars the next crucial step is to return rock samples so as to determine the nature of the rocks at the surface and their ages.

This progression of looking first at the gross features and then, on the basis of the global framework, studying the details of a planet, is quite the reverse of that of the study of our Earth. Geology came into being as a science with the increasing awareness of the necessity to examine rocks, their included fossils and associated landforms in the field. It is true that great philosophers, such as Descartes in the first half of the seventeenth century, concerned themselves with all-encompassing global theories about the origin of the Earth. Leibnitz interpreted the Earth as originating as a smooth, incandescent globe which, on cooling and contracting, developed a slag on the outside while water condensed from the atmosphere. As the crust cooled, great cavities developed, the roofs of which collapsed to form valleys, while the solid parts stood up as mountains. But such ideas were based on how things were thought to be rather than seen to be; it was not until people went out into the field and examined rocks themselves during the eighteenth century that the modern science of geology began to develop.

Early debates revolved around the origin of the rocks and fossils. Perhaps the most famous debate was the one in the late eighteenth century between those who believed that basalts were volcanic rocks erupted from volcanoes long since dormant, as advocated by the great French geologist Desmarest, and those who followed that strong-minded and overpowering mineralogist Werner who saw these rocks as precipitates from the oceans. As might be expected, the developing understanding of geology depended much on new observations made by travellers and explorers of the time. Travel in Europe probably played an important part, but the voyages of people like Cook in the mid-eighteenth century, and the studies of rocks, minerals and fossils during the great six-year expedition across the Russian Empire, initiated by the Empress Catherine, obviously provided much of the material on which ideas were to develop.

Early debates about the nature of specific areas of geology eventually led to a general acceptance of the basic principles of geology as formulated by that vigorous and observant Scottish physician James Hutton during the late eighteenth century. Hutton saw the Earth as being almost in a steady state. He postulated that processes of 'decay' were gradually destroying all the land by erosion. This 'decay' produced soil, and the detritus of the denuded land gravitated *via* rivers to the seas. The sediments which accumulated in the seas were then pushed up to form new mountain ranges which in turn became eroded again. He concluded that the Earth had a central fire which was responsible for the production of volcanoes, minerals and metallic veins. Thus he saw the Earth as being moulded by cyclical events, a concept that has become known as uniformitarianism, and a quite different proposition from that of the earlier catastrophists. Writers such as Charles Lyell in his *Principles of Geology*, the first volume of which was published in 1830, developed the ideas of Hutton, drawing heavily on examples of geological processes that could be seen to operate in many parts of the world.

It was perhaps Charles Lyell's work that finally established a tradition of geology which has continued to the present day. By the careful examination of small areas on Earth, a geological picture of the landmasses eventually emerged. However, this was still a very incomplete picture and only in the last few decades have we explored enough—although still very little—of the 70% of the Earth that lies under the oceans to enable a global theory, that of plate tectonics, to come into being. Before the theory of plate tectonics, geologists had recognized that powerful forces inside the Earth had greatly changed the face of the Earth through its history. It had been realized that compressive forces had thrust up the mountain chains producing intensely folded and faulted rocks, that earthquakes and volcanoes were mainly restricted to weaker belts in the Earth's crust where lava could rise from deeper levels more easily than elsewhere, and that constant movements produced earthquakes. Evidence had also been gathered to suggest that the continents had not always been in their present positions but had moved around the globe.

But all these ideas were based on studies of continental areas, which represent such a small part of the total surface of the Earth; studies of the ocean floors began to suggest that these different geological processes attributable to internal forces could be drawn together into the framework of a global history. Ocean studies indicated that there were areas of the ocean floor—for example, the mid-oceanic ridges—where new crustal material was being produced by volcanic activity. As new crust was formed the sea floor spread at perceptible speeds of a centimetre or so per year. Clearly, unless the Earth is getting larger, the new crust must be compensated for by the removal of crust elsewhere.

This removal takes place in areas such as the western coast of South America, where oceanic crust is being pushed down below the continent and causing uprise of mountains and volcanic activity. In earthquake belts, such as that of the San Andreas fault, the oceanic crust is sliding laterally against the continental crust without being either consumed or destroyed. Areas where one part of the crust is moving with respect to another are called mobile belts, easily recognized as those areas where there are high concentrations of earthquakes and volcanic activity.

Before these ideas developed it was known that the Earth consists of a number of layered shells of rock of different compositions. The outer layer, known as the crust, is made of rocks of relatively low density; below this is the mantle, which makes up the main bulk of the Earth's volume. At the centre of the Earth is the core, made up of nickel and iron.

The theory of plate tectonics sees the Earth in terms of the physical properties of the rocks. Floating on the outside is a thin layer of material that behaves essentially in a rigid way, at least on a short timescale. This shell, known as the lithosphere, extends down through the crust and into the upper mantle. Below this there are rocks of less rigid character which move in a plastic or ductile way; this region is known as the asthenosphere. In the theory of plate tectonics the lithosphere is seen to consist of many individual plates that float on the asthenosphere and are dragged around by, presumably, convection currents in the asthenosphere. On this model volcanism can be seen to be producing new crust as rock emerges from beneath, mountains are thrust up where plates push against one another, and great dislocations can occur where plates rub one against another. This general theory, now accepted by most geologists, provides a broad explanation for many of the phenomena produced by internal forces, and forms the beginning of an attempt to understand the Earth as a whole. It is perhaps not a coincidence that this global view of geology should come into being at a time when space exploration is dramatically emphasizing that the Earth is a planet and as such must be treated in the context of the Solar System.

A fundamental characteristic of geology is that the subject includes not only the study of the physics and chemistry of processes that produce rocks and landforms, but also history, to relate processes that occur at the present to those that occurred in the past. In this, geology has much in common with astronomy which, as well as observing the Universe as it is, tries to interpret its evolution through time. Thus the geological examination of a planet not only involves studies of processes such as meteoritic impact and volcanism, but also attempts to reconstruct the surface form of a planet and the nature of the processes operating at different stages in its history.

As we shall see, our most complete history of a planet other than Earth is that of the Moon. Careful geological mapping of the Moon from telescopic and spacecraft pictures developed what is known as a stratigraphy for the Moon; that is, a history of events based on the sequence of layered rocks exposed at the present surface. This history was a relative one, indicating the order of events; it was not until the rocks were brought back to Earth that actual dates could be assigned to these events. These studies emphasized the incredible age of the lunar surface. Briefly, we see that the heavily cratered surfaces on the Moon represent numerous impacts of asteroidal and smaller bodies into the lunar surface 4,000 million years ago and earlier. During this early history, volcanic events were almost certainly bringing material up from deep within the Moon to add to the surface crust, but much of the evidence for this volcanism was being destroyed by frequent impacts breaking up the rocks and destroying the landforms. The rate of bombardment appears to have slowed down rapidly just after 4,000 million years ago, and the dominant geological process then became volcanism producing vast floods of lavas now preserved as the *maria*. These catastrophic eruptions continued for about 1,000 million years; it is unlikely that much volcanic activity occurred after about 2,500-3,000 million years ago. Since that time little has happened to the Moon apart from occasional large impacts and a steady churning of the surface by small impacts. Thus, throughout much of geological time, the Moon has remained a relatively inactive body.

A similar situation is seen on Mercury, where most of the half of the planet photographed by Mariner 10 consists of an old cratered terrain. In some places these craters are buried by level tracts of relatively smooth-surfaced material giving extensive plains. As we shall see later in the book, the origin of these plains is not entirely clear; but it is possible that they are at least in part volcanic, corresponding to the *maria* on the Moon. It appears reasonable to interpret mercurian chronology in the same way as we do that of the Moon: that in Mercury's very early history it suffered intense bombardment, there followed a period of volcanism, since when very little has happened at the surface of the planet for the rest of its history.

When we examine Mars we see that this planet was also bombarded in its early history but, unlike the Moon and Mercury, the planet continued to evolve, producing large volcanic mountains, extensive floods of lava and, because Mars has an atmosphere, normal processes of erosion and deposition. Thus Mars has a more complicated geological history, although again much of the geological activity occurred relatively early in the planet's life, and at present martian geological processes are not operating at anything like the same rate as they are on Earth.

It is fascinating to realize that on the Moon, Mercury and Mars much of the recorded geological activity has been catastrophic; in the early history of the planet, and very occasionally in the later history, large impacts, some of which produced basin-like craters over 1300km across, must have drastically affected every part of the planet, showering the surface with high-velocity particles and shaking the ground with seismic waves; occasional but highly dramatic and extensive volcanic events took place spreading lavas over thousands of square kilometres; and on Mars other types of catastrophic events, such as flash floods, scoured the terrain. Not only were many of the events that formed the present landscape and surface rocks catastrophic, but the rate at which they occurred and their nature have changed with time so that, whereas in the very early history impact was the dominant process for all the planets, it gradually became a less important process and volcanism — together with, on Mars, other processes — became more important. It also appears that the internal activity of the planets has changed with time. Again, the planets were more active in their early history than they were later.

So when we look at the geology of another planet we see through history a series of catastrophies on a planetary body which is evolving in a unidirectional way with time. This is contrary to the normal tradition of geology based on our studies of Earth, where the philosophy of uniformitarianism has prevailed. This philosophy, summarized in the epigram 'the present is a key to the past', implies that what we see happening on the surface of the Earth today is representative of what has happened in the past, and that we can use our knowledge of the processes we see today to interpret the rocks and landforms of the past. This has been a very important concept in geology and has enabled great progress to be made in understanding the history of our Earth.

However, it should be remembered that until relatively recently it has been only the last 600 million years of Earth history that we have been able to study, by looking at rocks and fossils of the geological record. During this relatively short period of the Earth's history the processes operating may well have been similar to those we see today. Even so, some geologists believe that the rocks that are preserved were not always formed by the normal steady processes that we are familiar with, and that the geological record represents the unusual and often catastrophic events of the past. Certainly, when we start examining the rocks that represent the older history of the Earth, say before 2,500 million years ago, there is room to consider that not only were there catastrophic events but that conditions were quite different. The fact that other planets were heavily bombarded 4,000 million years ago and earlier strongly suggests that the Earth too suffered this bombardment, and our planetary studies at least suggest that the Earth has evolved in a unidirectional manner, the state of the Earth being quite different in its early history from what it is now.

It may well be that plate tectonics as we know it today did not operate during the Earth's early history. Although it is possible to find rocks, such as granite, typical of the later stages of Earth history in these earliest rocks, many of the structures are different from those associated with plate tectonics. This suggests that the thermal régime in early times was different from today. Also there is evidence that the Earth's atmosphere was different, possibly consisting of carbon dioxide.

Nevertheless, the Earth is in many ways quite different from the other planets that we have examined. The principle difference that we can observe at the surface as geologists is that the Earth, 4,600 million years after its birth, is still a highly active and mobile planet; with the Earth's atmosphere and free availability of water, erosion is taking place at a ferocious rate, wearing down the mountain ranges and depositing new material in basins; volcanoes are highly active, there often being as many as twenty or so volcanoes erupting at any one time, and some which erupt virtually all the time; earthquakes are much in evidence as tell-tale indications of the way in which the thin rigid plates that form the surface of the Earth are moving around at rates of many centimetres per year.

Why is the Earth so different? Clearly it has had a different thermal history and is still warm and active, whereas the Moon at least has cooled down relative to its condition 3,000-4,000 million years ago, when it could still erupt lavas at the surface. Why this is we do not know. At the time of writing, the Voyager I spacecraft is returning pictures of Jupiter's satellite Io which is a Moon-sized body that appears to have a very active surface; so not all small bodies have cooled down.

We have an admirable opportunity of investigating this subject further by the exploration of Venus, which has a similar density to that of the Earth, is almost the same size, and has an orbit around the Sun which is similar to that of the Earth. Consequently it is likely that Venus has a similar composition to that of our own planet and one might expect it to have had a similar history. One big observable difference is that Venus is shrouded by a dense atmosphere of carbon dioxide (perhaps the kind of atmosphere the Earth had in its early days). We are thus unable by conventional means directly to observe the surface of Venus. However, by radar we can look at its surface even from Earth, and so far there have been stimulating glimpses of features, some of which may be large craters, some volcanoes and others great rifts formed by faulting. We await with excitement the possibility of a spacecraft carrying a radar system to be put in orbit around Venus so that we can examine the surface in more detail and perhaps begin to understand why our own planet has the peculiar characteristics that we are familiar with.

Naturally, all our studies of other planets are coloured by our geological training and experience on Earth and, to any geologist used to solving problems of geology in the field, the study of other planets can often be frustrating because so much of the time we depend on data such as photographs obtained from an orbiting space-craft. Apart from at the few Apollo landing sites we have been unable to pick up rocks and bring them back to the laboratory for careful study, as would be the normal routine on Earth. On the other planets we rely much on what are called remote-sensing data, but it is worth remembering that, although we are now familiar with land areas on Earth, our knowledge of the evolution of the whole Earth depends on exploration of those areas covered by oceans, and we are now doing this using the very kind of techniques that we use on the other planets, those of remote sensing: the oceanographic ships that ply the Earth's surface are comparable in their rôle to the spaceships we have orbiting the planets; and the rare excursions in deep submersibles to the ocean floor use similar technology and are similar in character to our manned landings on the Moon. While in many ways we know considerably more about the Earth than we do about the planets — and this will be the case for many years to come — it must still be remembered that the major part of the Earth's surface still remains to be explored.

Geology is at an important turning point in its history; instead of looking at local systems of processes operating in relatively small areas of the Earth's crust through short periods of time, we may now begin to study the Earth as an integrated geological system and at the same time relate this to the evolution of other planets. Eventually it is hoped that a comprehensive theory will be produced to predict the conditions on any planet at any time given such parameters as the size and composition.

The planets have always been the domain of astronomy and to some extent still are, but the astronomers have suffered an invasion of geologists and other Earth scientists into their territory. Cohabitation between these widely differing disciplines has led to a realization that they have much in common. The astronomer is concerned with producing internally consistent models of the creation of a star from its infancy to its adulthood as a sun with accompanying planets orbiting around it. The Earth scientist, on the other hand, starts with the planets in their present state and attempts to work backwards in time to as near to the origin of the planet as he can get. Until very recently this was not very far, and the idea 'that it was the business of geology to discover the mode in which the Earth originated' was totally alien to geologists like Lyell and Hutton before him who declared that geology was in no way concerned 'with questions as to the origin of things'. Geologists' fear of things cosmological stemmed from their desire to establish geology as an observational science where one studied things that could be seen rather than speculated about things that could not. However, our planetary studies have now shown that we have the relics of ages not represented on Earth still preserved on planetary surfaces, and we may consider on a much sounder basis 'the origin of things'. The need to distinguish geology from cosmology does not perhaps have the same relevance to us as it did to geologists in the nineteenth century.

What can geological studies tell us about the origin of planets? Analyses of rocks brought back from planetary surfaces help in determining the composition of the planet as a whole, and this, in turn, allows us to start considering the composition of the material from which the planets were originally formed. However, as we have already discussed, the Earth is not homogeneous but is made up of a nickel-iron core surrounded by shells of silicate rock of different composition. Are all the

13

planets like this, or are some planets more homogeneous? Some would say that the Earth was originally formed with this layered structure; that during the process of accretion, when particles were coming together under gravity to make the planet, the early formed particles were metallic and only later were silicate particles added to the growing Earth. However, a more popular view is that the planets were formed from material that had a uniform composition and that in their early history they were homogeneous. There is the possibility that some, perhaps, have remained homogeneous, but this is certainly not the case with the Earth.

If we assume that a planet on formation is, for the most part, in a liquid state, then the elements will distribute themselves within the planet according to their affinity for different phases. These phases would be silicate melt, metallic melt, sulphide melt and a primordial atmosphere. The atmosphere would be controlled by the vapour pressures of the elements and stable compounds present in it, and would vary greatly in composition from one level to another, depending on temperature and pressure conditions at different levels. On the other hand, in the liquid material the elements would distribute themselves into crystal phases, depending on the amounts of different elements present and the affinities of these elements one with another. The denser phases, such as metallic iron, would separate under gravity to form a core while the lighter phases would float to the surface to form the mantle and crust. It is important to recognize that this process, differentiation, is controlled by the densities of the phases rather than by the atomic weights of the elements; thus heavy elements, such as uranium and thorium, will combine with the lighter phases and be carried to the surface of the planet rather than accumulate at great depth. The degree of differentiation that has occurred in a planet near the time of its formation is important in understanding its future history, especially with regard to such elements as uranium and thorium whose radioactive decay will be a heat-source for the planet, because if differentiation has gone to completion these elements will occur in the crust, whereas if no differentiation has taken place these elements will be distributed homogeneously throughout the planet.

From this simplified explanation of differentiation it becomes clear that studies of planetary geology can help in the quest to understand the origin of the planets and, indeed, the Solar System.

A study of the internal history of the planet through geology may also help us to understand how the planet was formed. One way of attempting to determine the starting temperature for a planet is to investigate its later thermal history and extrapolate backwards in time. Theoretical models can be produced that take into account all the processes known to occur within the planet, such as convection, heat-generation by radioactive elements, and so on; but such models can be greatly improved, firstly by knowing the present thermal régime of the planet and secondly by inferring earlier thermal régimes from our knowledge of the geological history; for example, in the case of the Moon any thermal model has to take into account the fact that volcanism occurred between 4,000 and 3,000 million years ago to produce the *maria* and that therefore the Moon was sufficiently hot several hundred kilometres below the surface to generate that magma.

The last few years, which have seen the birth and early growth of planetary geology and the beginning of exploration of the planets, have been exciting ones. In this book we hope to share with the reader a few of the interesting aspects of the surfaces of planets as we know them so far. Our aim is to illustrate what the different planets look like and to attempt an interpretation of the features that have been discovered. The reader will become aware that we do not have all the answers and there is still much work to be done by new scientists coming into the field.

The pictures we show are a small personal selection of the many tens of thousands of pictures that are now available and by use of which we may delve deeper into our study of how planets work. We have largely restricted ourselves to what we can see of planets using photographs, but it must be remembered that this is only part of the whole story, there being many other kinds of data that add to our knowledge: geophysical studies of the Moon, studies of lunar samples, meteorological studies of Mars, and so on. Nevertheless, the pictures are spectacular and we hope that this book will provide the reader with an introduction to planetary geology and a stimulus to learn more.

We hope that our professional colleagues will excuse us for not breaking up the text by giving bibliographic references to the writings of those who have created most of the ideas expressed by us throughout this book. Few of the ideas are our own, and those that are owe much to discussions and exchanges of ideas with our colleagues in this exciting field.

The Moon: Full Moon

On a clear night when the Moon is full, darker and lighter patches can be seen quite easily without a telescope. The darker patches came to be known as *maria* (singular *mare*), the Latin word for seas, for it seemed reasonable, by analogy with the Earth, that there should be seas on the Moon as well. As soon as telescopes were turned on the Moon it became obvious that the darker patches were not seas, merely large flatter areas of lower ground. There was no evidence of any water, clouds or atmosphere on the Moon, and altogether it appeared very different from its nearest neighbour in space, the Earth. In the four hundred years or so since the first telescopes were pointed at the Moon, no permanent change has ever been seen to take place on the surface, other than the slowly changing shadows as the Sun's rays progress across the lunar landscape month after month.

With the aid of a telescope, a number of types of feature can be distinguished. There are mountain ranges, such as the Apennines (the irregular arc-shaped feature marked A in the picture opposite) and also isolated mountain peaks, lower hills and ridges as well as the vast plains of the *maria*. But the most distinctive features are the craters, which vary greatly in size, and occur all over the lunar surface. These were assumed in the early days to be volcanic craters since, before the present century, volcanic craters were the only type known on the Earth. However, as early as 1665 an Englishman named Robert Hooke noted the likeness between lunar craters and craters formed by the impact of heavy objects dropped into soft pipe-clay. When meteorite falls became known in the early part of the nineteenth century, Gruithuisen, and later Gilbert, suggested that lunar craters might be the result of very large meteorites colliding with the Moon to form impact craters, thus beginning a major controversy over the origin of lunar craters which lasted until after the Apollo astronauts had landed on the Moon. The general consensus among scientists now is that both types of crater occur on the Moon, but that the impact craters are by far the more common. Ways of distinguishing the two types are discussed later.

The lighter areas of the Moon can be seen to be composed of hundreds of craters, crowding and overlapping each other to form a continuously cratered surface. These areas are called the highlands and are older than the lower-lying *mare* plains. Age relations between different rock units were first established by two Americans, Shoemaker and Hackman, in the early 1960s. They noted that the bright radial streaks called rays, radiating from craters such as Tycho (T) and Copernicus (C), overlie other lunar craters such as Eratosthenes (E) and are therefore younger. Eratosthenes itself has faint rays which lie on the Mare Imbrium surface, so the rays must be younger than the *mare*, whereas the *mare* material fills damaged craters such as Sinus Iridum (S) on the edge of the highland material, indicating that the highlands are older. These four time-divisions: Copernican, Eratosthenian, Imbrian and Pre-Imbrian became the major geological periods on the Moon (although Pre-Imbrian has now been replaced by Nectarian and Pre-Nectarian), and most of the Moon had been geologically mapped accordingly before any rock samples were brought back to Earth.

The Moon: Exploration

The Apollo missions to land men on the Moon, and the exploratory missions that preceded them were an exciting development in lunar science which took it from a primitive subject of little interest to a highly developed, fashionable and popular science in the span of a few years. Three Ranger craft crash-landed into the surface in 1964 and 1965, sending back pictures up to the last second showing lunar craters as small as a metre across. They were followed by five Lunar Orbiter craft in 1966 and 1967, the last two of which orbited the Moon in a polar orbit, taking pictures of virtually the whole surface, some of which appear in the following pages. At first everyone read their own theories and ideas into the new pictures, and controversy raged, but since that time many more missions and continued study have meant that a clear picture is slowly emerging. The Soviet Luna IX craft soft-landed on the Moon in 1966 and took pictures from the lunar surface, to be followed by seven US soft-landing Surveyor craft in 1966-68, five of which were successful, photographing small details on surface rocks and automatically scooping up and analysing lunar soil. Two Soviet Zond craft looped around the Moon in 1968 and returned safely to Earth.

But all these achievements were overshadowed by the Apollo missions, in which men first orbited the Moon at Christmas 1968, and finally landed on the surface on 20 July 1969. From that moment intense interest was centred on the Moon, particularly on the samples brought back to Earth, and they soon became better studied than any single set of rocks from the Earth. Those missions that landed on *mare* surfaces (Apollos 11, 12, 15 and 17) found the *maria* to be lava flows of a composition similar to basalt found on Earth, with ages ranging from 3,160 million years up to 3,960 million years. The rocks from the highlands were found to be intensely crushed and shocked rocks known as breccias, similar in many cases to breccias found in impact craters on Earth, with ages of about 3,900 to 4,000 million years.

The USSR meanwhile developed two automatic spacecraft, Luna 16 and Luna 20, which returned samples to Earth, and two Lunokhod craft which roamed the surface for 10km and 37km respectively. The later Apollo missions visited more adventurous sites, spent longer on the Moon, and had more sophisticated equipment such as the lunar roving vehicle with on-board cameras controlled from Earth by the umbrella-like antenna. Astronauts were able to reach a variety of interesting features such as the huge rocks thrown out from an impact crater (below and top left, Apollo 17) as well as take surface samples (top right). During the extensive traverses by the Apollo 17 crew, they encountered one of the most unusual lunar samples: orange soil caused by small orange beads formed in a volcanic fire-fountaining eruption 3,500 million years ago. A network of seismometers was set up which registered large numbers of tiny moonquakes. These gave clues to the deep internal structure of the Moon (see page 50), and a series of explosive charges fired on the Apollo 17 mission gave information on the near-surface structure in that region.

The Moon: Impact Craters

The dominant landforms on the Moon are impact craters; opposite is shown the crater Timocharis, a medium-sized crater 34km across. It is a comparatively young crater, so that the distinguishing features of impact craters are still fairly fresh, and clearly seen under the low-angle lighting. The crater itself is roughly circular in outline. This is in contrast with many volcanic craters, which are often rather elongated, as material is rarely erupted from a single point source. Outside the crater are the gently sloping, hummocky outer walls, which give way to roughly radial irregular ridges further out. This accumulation of material outside the crater is called the *ejecta*, and consists of material thrown out from the crater. It is naturally thickest near the crater, and thins steadily outwards, becoming discontinuous and patchy at a distance of about 1.5 crater diameters from the rim. Beyond this the secondary craters become prominent: these are craters formed by the impact of clots or blocks of material thrown out from the main crater (often called the primary crater to distinguish it from secondary ones). Secondary craters are usually fairly irregular in shape and tend to group in clusters or chains, similar to volcanic crater-chains, but they have a number of distinguishing features, described overleaf.

The mechanics of impact cratering have become better understood in recent years, thanks to small-scale experimental high-velocity impact studies by Gault and others, and also to the geological study of impact craters on Earth. The sequence of events in an impact is as follows:

Firstly the impacting body, which in the vicinity of the Moon is likely to be travelling at very high velocity, of the order of 20km per second, strikes the surface and begins compressing it downwards. Huge shock waves develop, and for a short time molten rock is squeezed out sideways from between the impacting body and the surface at extremely high velocities—much higher than the original speed of the impacting body. This matter is travelling so fast that it escapes the gravity field of the Moon altogether and disappears into space. Meanwhile the impacting body continues downwards into the lunar surface, becoming broken and compressed into a bowl shape, lining the crater which is beginning to form. Crater excavation is effected by the shockwaves initiated at the first moment of impact. These push the surface downwards and outwards at first, but the outward movement is soon deflected upwards, and material begins to be thrown out at an angle of about 40°, and a crater starts to be excavated. Velocities of ejected material during crater excavation are low compared with the original impact velocity, so this debris eventually falls back to the lunar surface. For a short while material continues to be thrown out, still at the same angle of about 40°, and the crater continues to grow, but velocities of the ejected particles rapidly decline to zero as the energy of the impact is expended, and excavation ceases. The total time to reach this stage varies with the size of the crater; for Timocharis it probably took rather less than a minute.

The Moon: Secondary Impact Craters

During the early stages of crater excavation, the material thrown out from the crater will eventually land great distances away to form secondary craters, but, as the velocity of throwout rapidly declines, it will fall nearer and nearer the crater, taking less and less time to land. The long chain of craters in the top left picture was caused by the impact of a string of clots thrown out from a large crater a long distance away. Note that, in this picture, wherever one crater overlaps another in the chain the one nearer the top usually overlaps the other. This indicates that the primary crater from which they came is in the direction of the bottom of the picture, because material thrown a greater distance from the crater will take longer in its flight, and land later than material landing nearer. Secondary craters further from the primary crater will therefore overlap the nearer ones.

The other pictures opposite show secondary craters and ejected material from the 95km-diameter crater Copernicus, at various distances from the edge of the crater. In each case, Copernicus is out of the picture to the bottom. Centre top shows a field of secondary craters 350km away from Copernicus, lying on Mare Imbrium. Because they have fallen close together in chains, and have frequently overlapped each other, most of these craters are irregular in shape, some of them being scarcely recognizable as craters. Ejecta thrown out from the secondary craters themselves can be seen as brighter ridged material trailing away from the crater groups in the direction of the top of the picture. It trails in this direction because the ejected material thrown out from Copernicus struck the surface at a low angle, throwing out material preferentially ahead of it in its direction of travel towards the top of the picture. Top right shows a crater chain 150km from Copernicus, with the craters in the chain so close together that it appears as a trough in the surface. Also visible is a herringbone pattern of faint ridges either side of the trough; this pattern is very characteristic of secondary craters, and helps to distinguish them from chains of volcanic craters, described later. The lower picture shows secondary craters about 100km from Copernicus, near the edge of the continuous ejecta. The herringbone pattern is evident again, but where it is associated with only one crater a single V-shaped ridge is formed instead of a series of Vs. These V-shaped ridges may be caused by bow-waves formed round ground obstacles in a dense cloud of gas, dust and ejected rock that streams away from the crater like a liquid. They may, on the other hand, be ejecta from the secondary craters falling asymetrically, as described overleaf. It has also been noted in laboratory tests that two almost simultaneous impacts occurring next to each other produce a V-shaped ridge between them strongly resembling those on the Moon. It is possible that lunar V-shaped ridges are the result of a combination of the above processes.

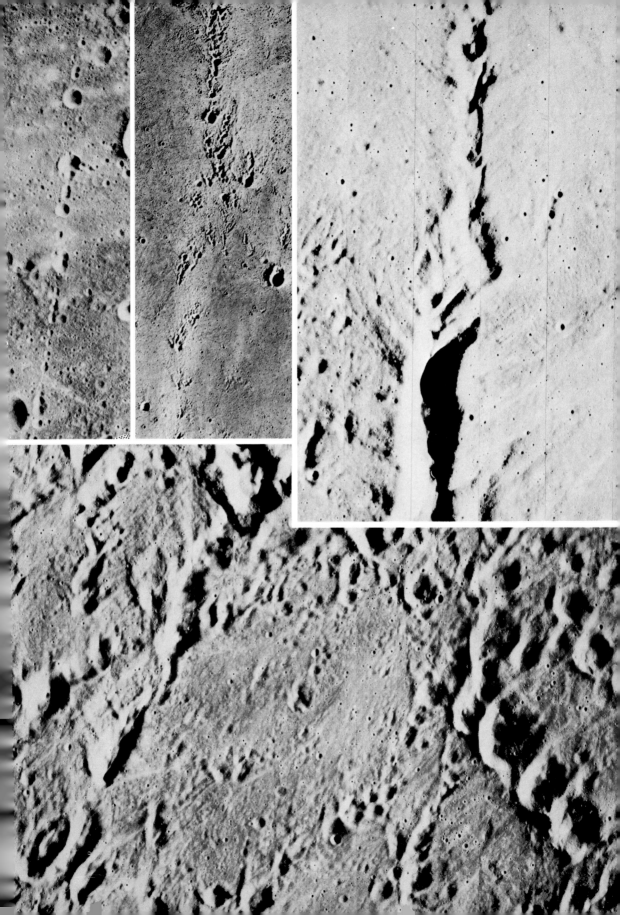

The Moon: Crater Rays

A moment's reflection will show that, as the Moon travels through space, not all bodies that it encounters will strike its surface vertically — indeed very few will do so. Most of them will strike the surface at varying angles from vertical down almost to the horizontal. However, even down to angles a few degrees from the horizontal, the resultant crater shape will differ little from a circle, because in all lunar craters the velocity of the impact is so great that the craters formed are much wider than the body which made them. This means that even if, in the early stages of crater formation, an obliquely impacting body strikes the Moon more sideways than downwards, the shockwave produced spreads out with equal velocity in all directions, becoming more circular the further out it goes, so that by the time the crater has enlarged to its full size it is virtually circular. The ejecta thrown out in the first moments of crater excavation, however, may not fall symmetrically about the crater. If the impacting body strikes the surface near the horizontal, it will be easier for material to be ejected ahead of the projectile. The resultant ejecta patterns are illustrated opposite. The top picture shows a crater on the farside of the Moon where the impacting body has fallen close to the vertical, and the ejecta, consisting of crushed and pulverized rock and therefore brighter than the surrounds, has fallen in fairly symmetrical radial rays about the point of impact. The lower left picture

shows the crater Proclus, which was caused by the impact of a body travelling towards the lower right corner of the picture. The upper left is therefore almost devoid of bright rays; had the impact been more oblique the sector without rays would have been still larger. Ray patterns like this have been exactly reproduced in the laboratory by firing very high-velocity projectiles at varying angles into beds of sand. The lower right picture shows, in the centre, the pair of craters Messier (right) and Messier A. These are a rare example of a very low angle, almost horizontal impact, where the impacting body, travelling from right to left, has skimmed off the surface once, losing much of its energy and velocity, and then collided with the surface again to form a second crater. Because the energy of the one impact has gone to form two craters, they have unusual features; both are highly elongated, the second one being itself a double crater. The ejecta patterns are distinctive: in the first crater all the throwout from the impact has been scattered sideways as a bicycle going through a puddle sprays water to either side. The resultant rays occur in a double-lobed pattern reminiscent of an eagle's wings. The rays from the second crater have fallen in a long double streamer ahead of the direction of impact. Again, double and multiple craters with ejecta distributions have been produced in the laboratory by impacting projectiles into sand at angles from 2° to 7° from the horizontal.

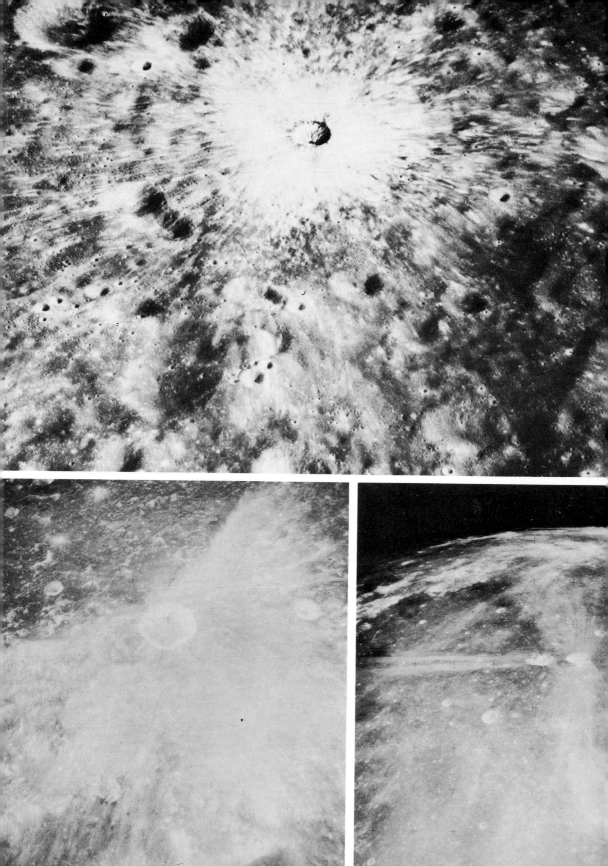

The Moon: Large Impact Craters

After the last of the ejecta has been thrown out, and the crater has ceased enlarging, there are still a number of events which may modify the appearance of the crater. For craters larger than about 15km diameter, the steep cliffs of the interior crater wall are too big to be supported at this angle, and large sections peel off and slump down to form step-like terraces. Opposite are the craters Aristarchus (top), 40km in diameter, and Tycho (bottom), 84km in diameter. The curved, slumped terraces can be clearly seen on the inside crater wall, the fractures associated with them running parallel to the crater rim.

In the last moments of crater excavation, there will be large quantities of broken rock, dust, and even some melted rock which fails to make it over the crater rim and gets trapped inside the crater walls. This will cascade down the inner walls as a debris flow (a mass of broken debris that appears to flow like a fluid) and spread out over the bottom of the crater. This may occur before, during and after the slumping phase. The flat floor resulting from this inward movement of debris is well seen in both craters opposite, and note that many very large blocks included in this movement are apparent on the floors of both craters.

While the slumping and debris flow is going on, another type of movement is occurring beneath the crater floor. For craters as big as Aristarchus and Tycho, the strength of the material beneath the crater floor is not great enough to maintain such a large hole, so large-scale movement begins upwards and towards the centre of the crater. The bottom of the crater rises to make the crater shallower, particularly at the centre, where a central hill is created. This series of events is similar to dropping a stone into water. At first a crater is formed, but the material strength of the water is not sufficient to maintain it, and the floor of the crater rebounds back to form a mound inside similar to the central peak of a lunar crater. (Since water has no material strength, both crater and central peak disappear fairly rapidly.)

The pictures opposite show part of the region immediately outside the crater rim. It is characteristically hummocky, with radial markings in some parts, and consists of the last ejecta to be thrown out before crater excavation stops. This ejecta had the lowest velocity, and so piled up closest to the crater without forming any secondary craters. The fine radial markings are grooves, debris flows and flows of rock melted by the heat generated in the impact; these are shown in more detail overleaf. Ejecta nearest the crater can be seen in cross-sections of experimental impacts to have been inverted relative to its pre-impact position. Horizontal strata appear to have been peeled back from the crater and laid upside down next to it.

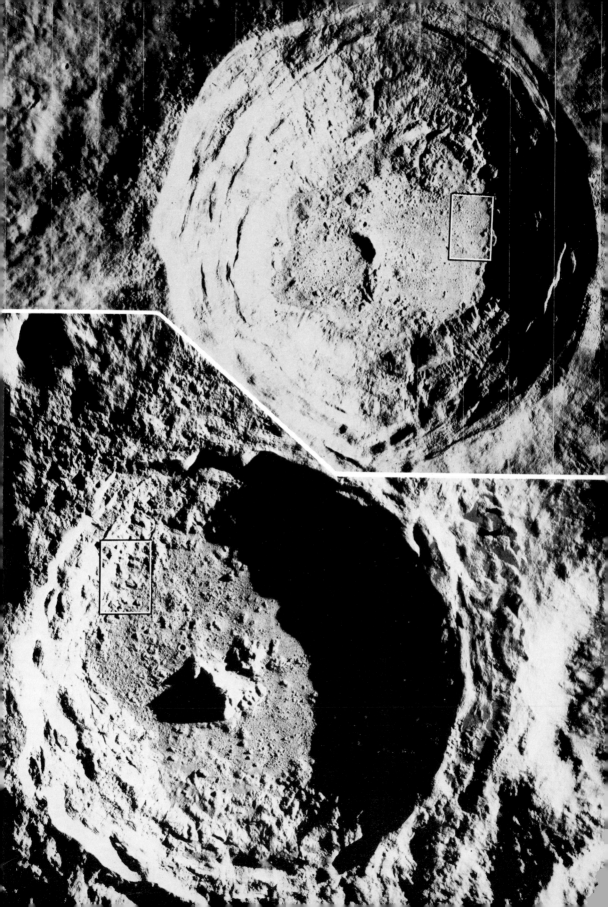

The Moon: Interior and Exterior Features of Large Impact Craters

The pictures opposite are close-ups of critical areas on large impact craters. The top left picture is an area 9km × 13km showing the terracing of the inner wall of the crater Tycho (the position of the picture is outlined on the previous page). The large terraces are prominent, as are a number of apparent flows (F), which have ponded to form flat areas (P) in some places, but generally run down towards the crater floor itself, which is at the bottom right of the picture. As explained on the previous page, it is probable that these are flows of rock melted in the heat generated by the impact, or debris flows, or a mixture of melt and fallback ejecta. They have clearly contributed to the crater fill forming the flat floor, but the flows marked in the picture must have occurred late in the impact process after terrace formation, for they run down valleys and pond in hollows in the already-formed terraces.

The top right picture shows an area 3.5km × 5.5km of the floor of Aristarchus, marked on the previous page. Note the cracks roughly parallel to the edge of the floor on the right of the picture. The texture of the floor resembles that of a lava lake, such as those found in Hawaii, apart from the difference in scale. The crack configurations resemble cooling cracks in a lava flow, and suggest that at least part of the crater fill was molten rock.

The bottom left picture shows an area 6km × 10km lying outside the northern wall of the crater Tycho. A prominent pair of thick flows with characteristic flow ridging are seen running up the centre, and a much thinner flow front (f-f) to the right of the picture. There are also some flat ponds (P) in hollows near to the top of the picture. These flows, which strongly resemble viscous volcanic lava flows, were thought by many to be just that, and even today some who accept that Tycho and Aristarchus are impact craters do allow the possibility that these flows may be from later volcanic eruptions long after the impact event. The lack of visible vents, however, can more easily be explained if these flows are molten or partially molten rock heated and ejected by the impact. The ponds could be much more fluid examples, filling hollows in the ground. It has been shown that these ponds tend to occur on the downrange side of an oblique impact; on Tycho they concentrate outside the crater's east rim, showing that the impacting body was travelling east when it struck the surface.

The bottom right picture shows a large debris flow (the flow front is arrowed) associated with the very large crater Tsiolkovskiij 185km in diameter, on the far-side of the Moon. The rim of the crater is seen in the left of the picture. This is a debris flow on a far larger scale than the small features found on the inside and outside walls of Tycho and Aristarchus — the whole outer rim of Tsiolkovskiij appears to have slumped and travelled outwards in this section. The area shown in this picture measures 110km × 140km.

The Moon: Small Impact Craters

The picture opposite is a panorama of the small crater Shorty, 100m across, taken during the Apollo 17 traverse. Note the many smashed and broken rocks that have fallen back into the crater, and also the rocks outcropping at the rim, such as the large one behind the lunar rover at left. Small craters are relatively simple features, lacking the complicating modifications of larger ones. The material strength of the surface is capable of supporting a hole of this size, so there is no central peak. Similarly, no slumping occurs down the inner walls, so no terraces form. The resultant crater is usually a simple bowl shape, with a circular rim; only the presence of broken rocks disturbs the symmetry. The outer walls are hummocky, however, and outside these are the ejecta, secondary craters and rays, although they have a characteristically different appearance from those associated with larger craters. Small, bowl-shaped craters like this can be seen peppering the whole lunar surface. Other pictures in this section show them; they are found on all types of lunar terrain.

Such small impact craters have been used to date different areas of the Moon. It was shown on page 16 how features were dated relative to each other by superposition relations; clearly a feature which partly obscures another feature by overlying it must have been formed later. Using numbers of small impact craters, relative dating of two surfaces can be done without them being in contact. For relative dating, it suffices to count the numbers of craters in a given size range on equal areas of the two surfaces to be dated. Assuming an uninterrupted rain of impacts on the Moon, clearly the older a surface is, the more impact craters it will have upon it. Aristarchus and Tycho (previous page) are very young craters and there are very few impact craters visible on them, whereas the craters Archimedes and Fra Mauro (page 35) are much older and have many smaller craters scattered over them.

By counting the number of craters in a given area around each of the Apollo sites, the relation between numbers of craters and the dates measured from the rocks was established. Ideally, this relation could be used to date any area on the Moon, but in practice it is complicated by secondary craters, which in a region such as that illustrated on the lower half of page 21 might be confused with primary craters and give an age much older than the true one. In hilly areas, on the other hand, small landslides and other downward movement will tend to destroy small craters after they are formed, so that too young an age will be assigned. Also, on old surfaces the small craters themselves will eventually be filled in and eroded. The crater in the picture opposite is already starting to be eroded — the fresh angular rocks that once lined its rim have been worn down, and the floor is gradually filling up with debris from other impacts nearby.

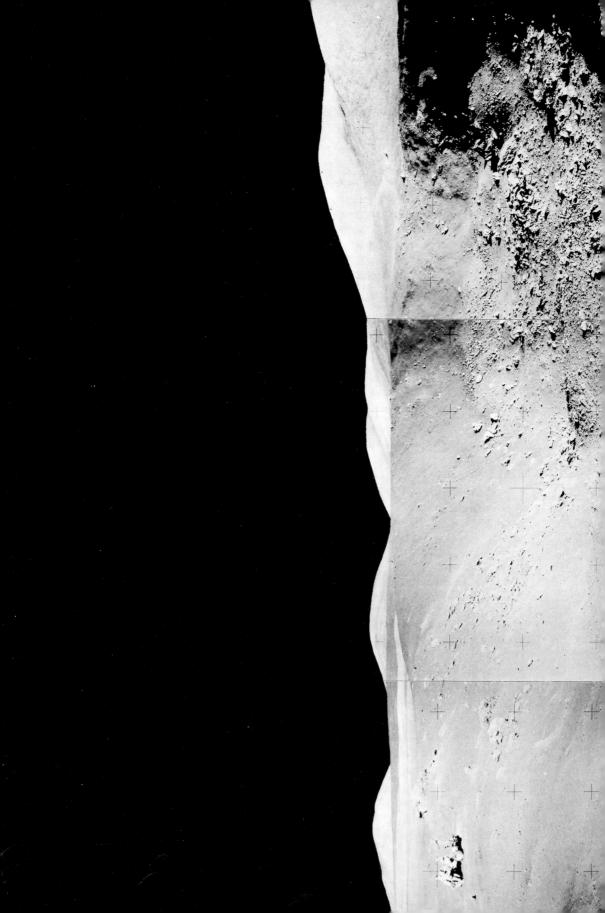

The Moon: Comparison with Impact and Explosion Craters on Earth

One of the main difficulties in studying lunar craters, until recently, was the lack of knowledge about what a fresh impact crater looks like. By far the best preserved impact crater on Earth is Meteor Crater, Arizona (top right), only a few tens of thousands of years old, but even it has suffered considerable erosion as manifested by the water gullies down the inside, and the texture of the ejecta and secondary craters outside the rim appears almost entirely smoothed over by wind and water when compared with the lunar crater Diophantus (top left), which is probably approaching a hundred thousand times older. There are only broad similarities between the two, like the hummocky rim. Another problem is that of scale. Diophantus is 18km across whereas Meteor Crater is only just over 1km in diameter; as explained on page 26, craters more than 15km across normally have some features not found in smaller craters. There are many larger impact craters that have been discovered on the Earth in recent years, but usually the rim has long ago been eroded down and the crater filled in, and often all that remains are the roots of the crater. This has been very useful in determining the subsurface structure of impact craters, but has revealed little of their original surface features to compare with lunar craters.

For this reason scientists have sometimes been forced to use bomb craters for comparison. The heavily cratered battlefields of northern France in the First World War frequently suggested a lunar landscape to ordinary soldiers, as well as scientists; in fact, the bomb cratering of the First and Second World Wars together probably gave the major impetus to the impact theory of lunar cratering as summarized by the American Ralph Baldwin in the immediate post-war years. When the very large craters caused by nuclear explosions such as the Sedan crater (bottom) began to be studied, many similarities to lunar craters were found. The hummocky rim, the roughly radially arranged ejecta further out, and beyond that the clusters of secondary craters with their own ejecta falling away from the primary crater in ray-like streaks — all resemble closely the arrangement around lunar craters, bearing in mind the probable differences in surface characteristics, particularly the presence of water, and also the gross difference in scale. But despite the similarities of the final product, it must never be forgotten that there are distinct differences between impact-cratering and bomb-cratering mechanics.

The present fairly detailed knowledge of what happens during an impact event, and what the crater should look like afterwards, has had to be pieced together from essentially three sources of information: (1) the laboratory experiments described earlier which produce true impact craters but are too small to show the characteristic diagnostic features of larger craters; (2) the large natural impact craters on Earth which give much information on the process of cratering but whose top surface has been eroded away; and (3) high-energy and nuclear explosion craters, which are fresh enough to show detailed surface features but are not true impact craters.

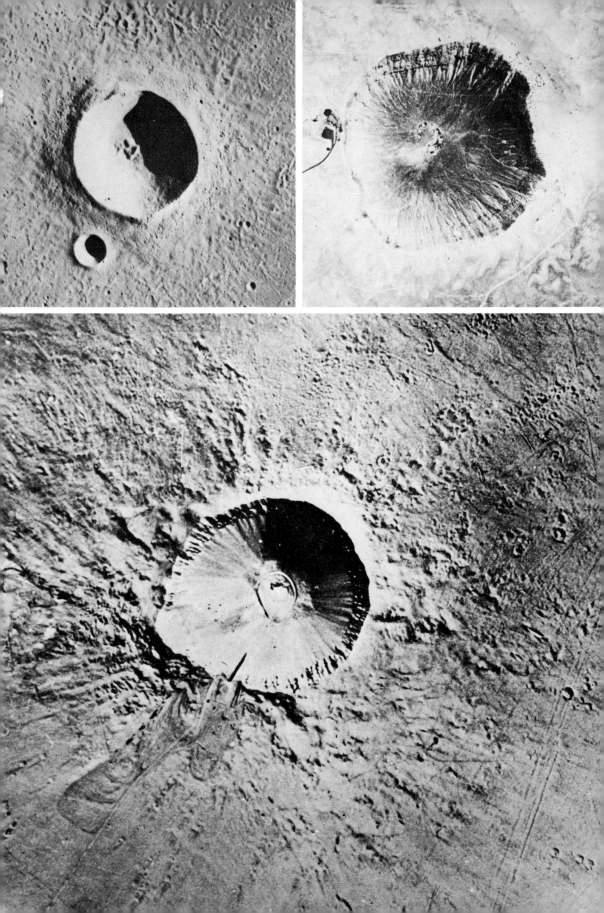

The Moon: Erosion and Burial of Impact Craters

It is not always appreciated that the Moon, like the Earth, has its own erosive processes which slowly wear down craters, mountains and other surface features. Nearly all erosion on Earth is accomplished by the action of water, ice and, to a lesser extent, wind. Since all three of these processes are absent on the Moon, what is there left? The answer is certainly very little, and it must never be forgotten that the amount of erosion effected on the Earth in thousands of years would take hundreds of millions of years on the Moon. Two processes are actually accentuated on the Moon by the lack of an atmosphere. One of these is cosmic-ray bombardment, which may break down surface rocks on a very small scale; the other is meteoritic bombardment. Cosmic rays are prevented from getting through to the Earth's surface by the atmosphere, and similarly all meteors below a certain size are heated by friction with the Earth's atmosphere until they melt or shatter, eventually landing gently on the surface as fine dust. On the Moon these same particles will strike the lunar surface at their full orbital speed of several kilometres per second, causing craters of varying dimensions. (The particles themselves are known as meteoroids while they are in space; the phenomenon of a meteoroid encountering a planet's atmosphere is known as a meteor; a meteoroid which survives this passage to reach the ground is called a meteorite.) The effect of millions of years of continued bombardment by meteoroids has meant that the lunar surface has undergone a continuous breaking up and churning over of material, sometimes appropriately called meteoritic 'gardening'. The resulting layer of broken rocks and finer dust at the surface is called the regolith, which may be several metres thick. The bombardment causes a smoothing over of all surface features, so that large impact craters gradually lose their secondary craters, V-shaped features and other identifying marks. This process is especially effective on slopes, where an impact may start an avalanche of rocks. Moonquakes, often caused by large impacts, may initiate erosion by starting landslides.

The effects of some of these processes are seen in the craters opposite. The large craters Fra Mauro and Bonpland (middle left) are very old, and the effects of heavy meteoritic bombardment can be seen on their floors, while the broken rims show only the bare outline of a circle. In addition they are criss-crossed with fissures, and the left half of the top crater (Fra Mauro) is covered with ejecta from a much larger impact. It is hard to imagine that these craters once resembled Aristarchus and Tycho (page 27). Top right shows a similar ancient crater, but with the nearside down-faulted and subsequently flooded by lava, so that all that remains is a semicircle. The lower right picture shows the crater Archimedes (80km), which has had its central peak and interior features submerged beneath lava flows, and most of its lower outside slopes.

The Moon: Faulting

From the earliest days of telescopic observation, straightish lines were noted at many places on the lunar surface. They were named rilles — an early spelling of rills, meaning tiny watercourses: the different spellings are maintained to avoid confusion. The largest of them is the Sirsalis rille (left), which is 430km long and averages about 3.5km across; smaller ones are found down to the limits of visibility. Rilles occur on all types of terrain, although they are concentrated round the edges of the *maria*, and are comparatively rare on the farside of the Moon. The two sides of the rille are two faults formed as the land to left and right moved apart; the central section of the rille dropped down between them. Rilles tend to broaden when they cross higher ground — note how the Sirsalis rille is broader where it crosses the rim of an older crater near the bottom of the picture than it is on the floor of the crater. This indicates that the two bounding faults slope towards the centre of the rille. These structures are well known on Earth, and are generally called graben. Single normal faults, where the ground on one side of the fault drops relative to that on the other, are far rarer on the Moon. The best known example is the Straight Wall (second from left) which is 115km long and about 400m high. The second picture from the right shows a series of *en echelon* graben, stretching for 250km. Such patterns are often associated with tensional movement, and sometimes with strike-slip movement, where the land on one side of a fault moves horizontally in a direction parallel to the fault. The right-hand picture shows the Hyginus rille, bisecting the 9km crater Hyginus near the bottom. This is a normal graben, apart from the series of craters along it, which includes Hyginus itself. Clearly these are not impact craters, for it would be too much of a coincidence for so many impacts to have occurred on the rille. Also they differ from impact craters in having no raised rims; they are simply holes in the ground. This suggests that they are collapse craters, formed when something was withdrawn beneath them; such craters are common in volcanic provinces on Earth. Withdrawal of the magma beneath the fissure may have caused these collapse craters to form over specific vents along the fissure.

The distribution of faults and graben over the Moon shows an interesting pattern, for there are more of them in three preferred directions: northwest-southeast, northeast-southwest, and north-south. This pattern is known as the lunar grid pattern, and is apparent not only in faults and graben but also in other features such as crater rims, which may be slightly polygonal or contain fractures in the above preferred directions. It is clear that the grid pattern is a fundamental pattern in the lunar crust, possibly induced by the tidal attraction of the Earth, or perhaps caused by the Imbrium basin impact described later, and any fractures that develop are liable to trend in one of these directions.

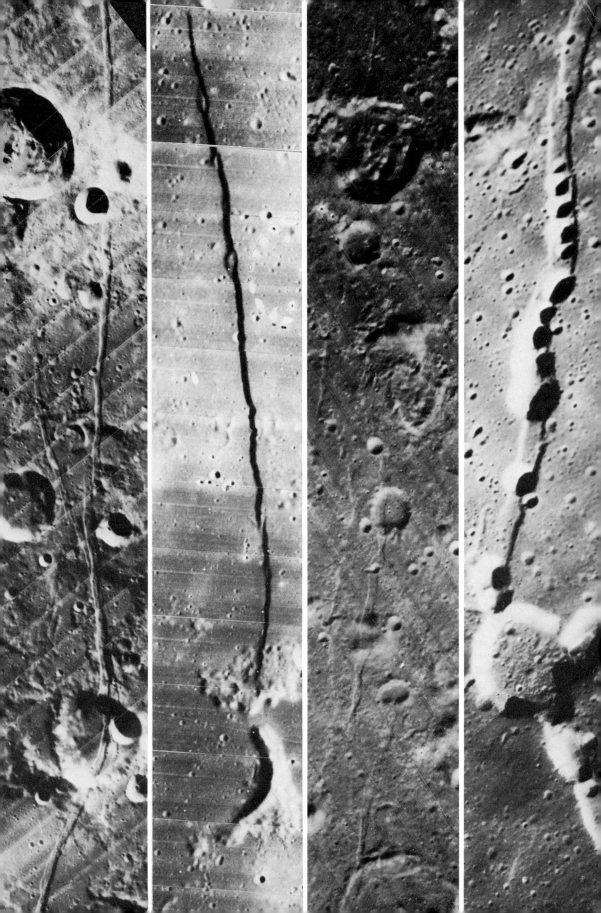

The Moon: The Maria

The extensive plains of the *maria* appear flat and featureless under high illumination, but under a low sun angle (opposite) the myriads of tiny craters are thrown into exaggerated relief by their long shadows, and many unsuspected low ridges appear. Even smaller-scale height variations are visible in the left half of the half submerged crater Letronne (120km) where the sunlight falls at the lowest angle: the *mare* surface here takes on a continuously undulating appearance, but the vertical extent of the undulations is only of the order of a few metres or less. This poses a problem: if the *maria* really are lavas, then under these lighting conditions the lava flow fronts should be clearly visible, but they are not. Furthermore, in the area of *mare* surface shown opposite—some 27,000km² — there are plenty of impact craters with their circular rims and low outer slopes, but no volcanic cones, nor signs of flow to indicate where the lava came from. This problem was solved soon after the first samples came back from the Moon. When rocks of the same composition as the lunar ones were melted down, they were found to be far more fluid than any lavas found on Earth. Their viscosity was roughly equivalent to engine oil so that, instead of flowing for short distances and building up substantial lava fronts like lavas on Earth, they were able to flow great distances before solidifying, and to fill in hollows almost like water leaving no visible fronts or flow features to indicate where one flow ends and another begins. There are clues in the picture opposite, however, that there were successive flows of lava, rather than just one big flow. The 13km crater Flamsteed A (arrowed) has been completely surrounded by lava, indicating that the crater is older than the lava. Lava has covered its secondary craters and hummocky ejecta except the immediate outer walls, showing that the lava covering is fairly thin here. Yet the much larger crater Letronne has had its walls completely submerged in this region, indicating a much greater thickness of lava. Flamsteed A must, therefore, have been formed after most of these lavas had been laid down, but before the final flow or flows which surrounded it.

From the foregoing discussion of the lava viscosity, it is clear that *mare* surfaces should be smooth and flat, yet wrinkled ridges dominate most of the *maria* regions, frequently paralleling the edges of craters, mountains, and the edge of the *mare* itself, but often apparently bearing no relation to them. Note opposite that the crater Letronne has had its northern wall obscured by lavas that have subsequently flooded into the crater, covering the floor and leaving only the very highest peaks of the central mound still visible, yet the missing section of northern wall is almost replaced by *mare* ridges. Flamsteed A, however, seems to have had no effect on the directions of *mare* ridges; the small ridges are interrupted by the crater, but continue the other side as if it were not there.

The Moon: Mare Ridges

Further examples of *mare* ridges are shown opposite. Basically, they can be divided into two types (top left); the broad, shallow-sloping ridges (B) and the narrow, steep-sided ridges (N). The broad ridges are typically about 5km wide, with slopes of about 3°, and do not interrupt or overlap craters or other features superimposed on the *mare* surface. The narrow ridges are frequently superimposed on the broader ridges, usually running along one side or the other. The situation in the top left picture, where two broad ridges parallel to one another have narrower ridges on opposite sides, is a common one, and can be seen in both the lower pictures. The narrower ridges are typically less than 1km wide, with slopes of about 15°; unlike the broad ridges they sometimes overlap craters on the *mare* surface, indicating that they were formed later, and also that they were composed of a substance able to flow. It is clear from the disposition of the *mare* ridges that they were formed by folding and buckling of the surface. Similar forms, particularly the *en echelon* formations, can be produced by moving the hand about on a tablecloth. The *mare* lavas would have taken some time to cool, and a surface skin of solid rock would have developed, while the lava beneath was still molten. Movement of this solid skin as the molten lava beneath cooled and contracted would have caused it to break up into separate plates, and, as the boundaries between plates solidified again, further plate movement would have produced folding and buckling along those boundaries. The broad *mare* ridges are clearly folds, whereas the narrow ridges are probably later lavas that have extruded through cracks to the surface, and having cooled off considerably were more viscous than when they were originally erupted.

The association of *mare* ridges and fractures is clear from the lower left picture, where the two parallel one another in this 115km × 115km area at the edge of the Mare Serenitatis. *Mare* ridge configurations are frequently influenced by features beneath the *mare* surface. At top right the ring-shaped feature R (6km across) indicates the site of a buried impact crater, and on the previous page the buried walls of Letronne are marked by a curved series of ridges. Clearly the subsurface topography will influence where the edges of the plates, and consequently the ridges, will form. Furthermore, any slight movements along the circular fractures that surround impact craters will cause buckling at the surface.

Not all the *mare* ridge formation takes place while the lava beneath is still molten, however. The lower right picture (115km × 155km) shows ridges near the edge of the Mare Serenitatis. The *mare* surface near the bottom of the picture has many more impact craters upon it and is therefore considerably older; yet the ridges in the newer *mare* surface above continue into the older *mare* surface without interruption. The lava in the older *mare* must have cooled in the great span of time implied by the difference in numbers of craters on its surface, so the ridges here must have formed by folding of solidified lava. It is possible that intrusion of lavas in sheets below this older surface has provided the lubrication for surface movement.

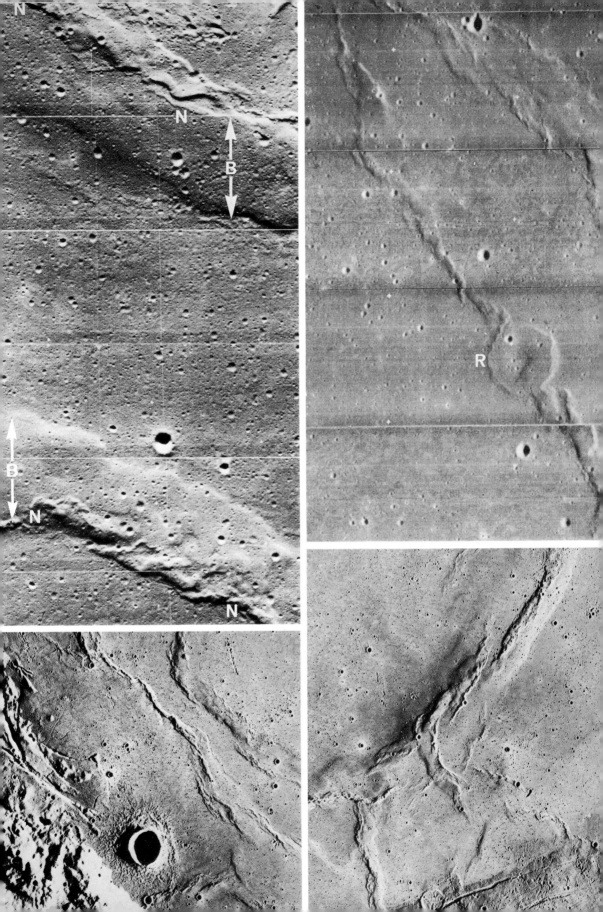

The Moon: Imbrium Lava Flows

There is one exceptional area in the *maria* where the marial lavas were viscous enough to form lava fronts. This area is shown opposite, and lies in the southwest part of the Mare Imbrium. The information this area provides about the mechanism of eruption of *mare* lavas contains some surprises. Firstly the great size of the individual flows is without parallel on Earth. They have travelled distances of up to 600km, and the flow fronts average about 30m in height, yet the slopes down which they flowed were 1 in 100 at the most. It has been found from studies of volcanoes on Earth that the distance a lava travels, though dependent on viscosity and slope, is largely determined by the rate of eruption of lavas at the vent. In the case of the lavas opposite, a very high eruption rate is probable, again much higher than any known on Earth. Another peculiarity of these lavas is the problem of where they come from. It is clear from the picture opposite that the lavas flowed from the bottom towards the top, but when they are traced back towards their source (out of the bottom of the picture) no volcanic cone is found from which they could have issued. This implies that they issued quietly and rapidly without any explosive activity, from an area on the southern borders of the Mare Imbrium. Taking this as typical behaviour of all marial lavas explains why no visible volcanic structures are unambiguously related to the large marial lavas. The most likely candidates for volcanic vents are long fissures such as are frequently found at the source of large flows on Earth. In the case of the normal very fluid *mare* lavas, ponding over the vent may drown the fissure; other possible candidates for vents are discussed later.

The composition of *mare* lavas has been analysed from rocks collected at five sampling areas on the Moon. Although they are similar to basalt lavas on Earth, there are some important differences. For a start, the lunar lavas contain no water or hydrous minerals, suggesting that water has not been present in any large quantities for the last 4,000 million years on the Moon. They contain more titanium, although the amount varies between the various sites at which samples were taken, and more iron too; not surprisingly the iron oxides are usually in a less oxidized form than on Earth. Sodium and potassium are very low in the lunar lavas, and this, together with the lack of water, suggests that the Moon was extremely hot at some time in its history, boiling off these elements into space. The familiar minerals found in basalt — pyroxene, olivine, plagioclase, ilmenite and spinel — are all present, but there are some new minerals not found on Earth. One of these, pyroxferroite, is similar to pyroxene but has more iron. Another new mineral, composed of iron or magnesium and titanium oxides, was called armalcolite after the Apollo astronauts, *Arm*strong, *Al*drin and *Col*lins. It has since been found in South Africa. The sizes and shapes of the crystals in the lavas indicate that they cooled relatively rapidly — in less than a few years. This is additional evidence that the *maria* are composed of several thin flows — less than a few tens of metres — for had they been a few thick flows they would have taken hundreds or even thousands of years to cool.

The Moon: Lunar Volcanic Craters

Lunar volcanic craters are rare and very small. There are no large shield volcanoes like Etna or those in Hawaii, and there is certainly nothing resembling the giant shield volcanoes of Mars (described later), but in a few areas a variety of volcanic vents are found, some of which are illustrated opposite. At the top left are three horseshoe-shaped cones, marked P, the largest being 2km across. They are easily distinguishable from the impact craters (I) by their wide outer slopes as well as the breach through which lava has been erupted. They stand out as positive constructions — hills as opposed to the holes of the impact craters — indicating that something has been erupted. At top right is a series of similar craters, but this time aligned in a row, with a prominent fissure extending from them. This appears to be a small fissure eruption, which has built up explosive cinder cones, but where no lava has been emitted. Middle left shows a very different kind of crater (C), which is too elongated for an impact crater and lacks visible outer walls. It measures 5km × 3km and contains a pronounced terrace, double in parts, on the inside wall, and generally resembles the collapse pits or calderas normally found on basaltic volcanoes on Earth. The only difference is that, on Earth, such pits are usually found at the summits of gently sloping volcanoes. It may be that on the Moon the lava was too fluid to build up any edifice at all, but spread out rapidly to form the level *mare* plains round about. Note the difference between this volcanic collapse crater, which has no outer slopes, and the secondary

impact craters (S) which are more irregular and have the shallow outer slopes and ridges resembling V features typical of secondary craters. At middle right is another fissure eruption, at F, and also a 10km-wide lunar dome at D, with a summit crater (C). This is probably of similar origin to the caldera at middle left, but in this case the lavas were slightly less fluid, and have built up the small circular swelling round about. At bottom left is a volcanic complex near the Mare Orientale, with a prominent fissure (f) with pyroclastic cones on it. Lavas from this fissure have fronts at L, and a circular dome of lavas (d) is visible at left without a summit caldera, whereas the lava front (L) at lower right has a possible caldera (C) associated with it. The picture is 20km from left to right. The bottom right picture shows a 4km-wide steep dome superimposed on a shallow swelling resembling a normal lunar dome. The two probably have a similar origin of lava extrusion, but in the case of the steep dome the lava is more viscous, and forms a steep-sided irregular flow. Both this and the cones at top left are from the Marius Hills, the largest area of volcanic vents on the Moon, being over 270km across. There is no evidence, however, to suggest that this area or any of the features illustrated opposite are sources for the major lava flooding of the *maria*. Most probably they are the sites of minor late eruptions, in some cases perhaps extrusion of still-molten lava from under a solid crust formed over the cooling lava of a major *mare*-forming eruption.

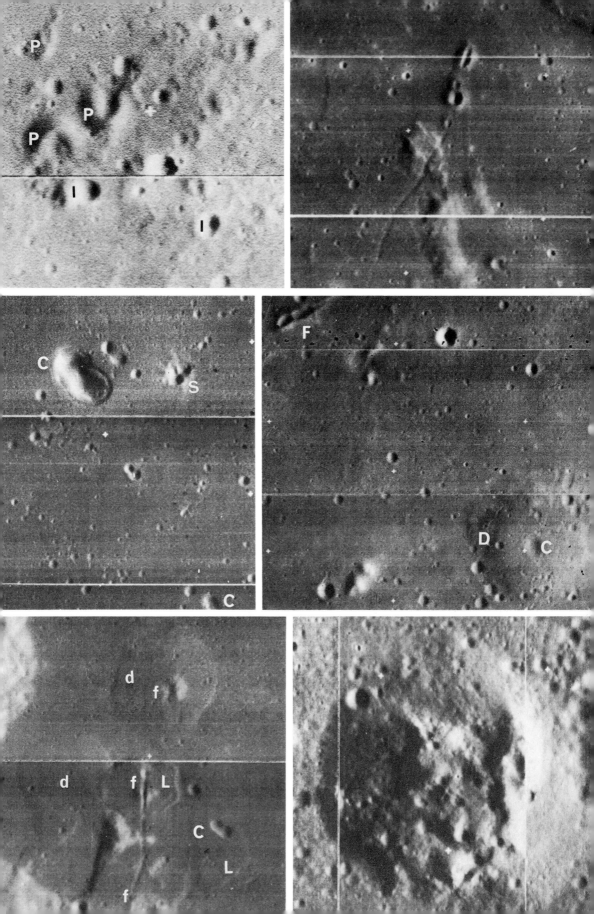

The Moon: Sinuous Rilles

One of the most curious types of feature photographed by the Orbiter probes were the sinuous rilles, which differ from normal rilles in having meandering bends like a river.

They had been first noted much earlier, at the turn of the century, but their discovery in large numbers gave rise to great excitement and interest, for it was thought for a time that they might have been formed by running water, and that their concentration in some areas — e.g., near the edge of the Aristarchus Plateau (opposite) — indicated larger amounts of water in such regions.

The lack of water in the lunar samples, however, demolished this theory, and besides this the sinuous rilles show a number of differences from rivers. Most of them start in a rimless depression, and the rille begins broad and deep, but both the depth and the width tend to decrease towards the other end, although occasionally the depth may vary abruptly, contrary to this trend, and in some cases a complete break may occur in the rille.

The start of a rille is at a higher altitude than the end, although the downhill slope is not always continuous throughout the rille; but, where there are uphill sections, they can nearly always be shown to be associated with later up-arching. In addition, natural water channels on Earth are not so deep in relation to their width, and show other signs of flowage and depositional features at the downhill end.

But other characteristics of sinuous rilles are similar to those of a flowing liquid. They are diverted by obstacles, hug the edge of high ground, and follow pre-existing depressions. For example, in the picture opposite, there is a series of fractures running left-right to the right of the crater Krieger, the largest in the picture. Some of the sinuous rilles which cross the fractures are abruptly diverted along them, becoming straight for that portion of their length, before being again abruptly diverted on their previous winding course.

If the liquid is not water, then what is it? The answer is almost certainly molten lava. Lava channels running away from eruption craters are well known on Earth, as shown overleaf, and share a number of the peculiar characteristics of sinuous rilles. Sinuous rilles are fairly common on the Moon, and the craters at their heads may well be major sources of the marial lavas, but other observations suggest that they are not the only sources. The large flows in the Mare Imbrium described on page 42 are not fed by sinuous rilles, and, though these are not typical of major *mare* flows, there are other large areas of *mare* which have no sinuous rilles which could have fed them, so quiet but rapid emission from fissures remains an important mechanism of eruption for the marial lavas.

The Moon: Lava Channels and Tubes

The two pictures on the right show lava channels in Idaho (top) and Hawaii (bottom). They have general characteristics similar to sinuous rilles, commencing in a crater which may be circular (top) or elongate (bottom). The width and depth tend to decrease away from the source crater, but close examination shows that there are small abrupt increases and decreases in depth and width along the whole length of the channel, and there are frequent breaks in it. Lava channels like those in the top right picture have actually been observed during formation. As the molten lava from the source crater or fissure flows away downhill, a solidified crust forms on the surface due to cooling. This crust thickens in areas where the lava is moving slowly, and it is only where the lava is moving rapidly that it remains fluid. These areas of rapid movement soon settle into well defined channels of lava. Sometimes a cooling crust may develop on the surface of the flow so that, for part of its length, the channel may be roofed over and flow underground. An underground lava channel is known as a lava tube. The formation of this crust is assisted by 'rafts' of solidified lava floating on the flow surface and getting stuck at narrower or more sinuous parts of the channel. Another mechanism of roofing over begins with the formation of levees (higher banks) on either side of the flow. These are initiated by small repeated overflows from the channel, which rapidly solidify and build up walls on either side, which may meet over the top of the channel to form a tube. Once a channel has been roofed over in any of these ways, the presence of a roof provides insulation for the lava underneath, so that it remains hot and therefore fluid, and can reach the advancing front more efficiently. Once the eruption stops, the lava will flow out of the tube sections to leave hollow underground caves, which may subsequently collapse. It is clear from the two pictures on the right, particularly the lower one, that for a large part of their length, these channels are not lava channels proper, but collapsed lava tubes. The collapsed roof can in some places be seen lying in the bottom of the tube, and the breaks that occur in these features are sections where the roof has not collapsed. Despite this resemblance to lunar sinuous rilles, there remain some distinctive differences, not least of which is the difference in size. The top left picture shows the largest lunar sinuous rille — Schroeter's Valley. It is over 150km long, 4km to 6km wide, 500m deep, and contains another sinuous rille inside it. No lava channel on Earth approaches anywhere near these dimensions; the one in the top right picture is only 3.5km long. Another difference is that lava channels and tubes tend to lie on the summits of ridges. This is due to the repeated overflowing of the lava channels which build up levees. On the Moon there is no corresponding rise adjacent to the sinuous rilles. Also, although lava tubes and channels do wander and meander slightly (right) they never show the highly sinuous snake-like meanderings of some of the lunar sinuous rilles, like the one inside Schroeter's Valley (bottom left).

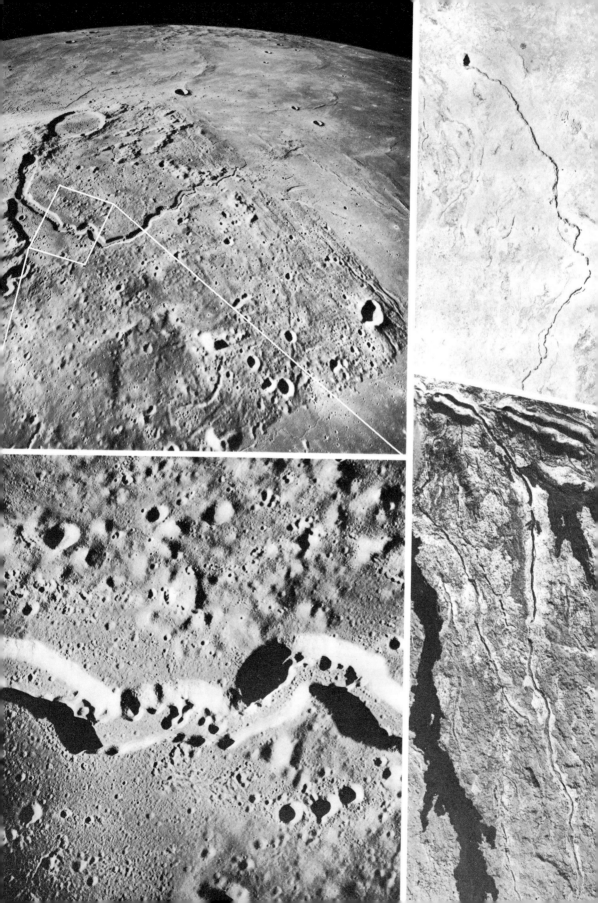

The Moon: Hadley Rille

All the differences between lava channels on Earth and the sinuous rilles on the Moon may be explained by the low viscosity (which is probably due to the higher titanium and iron content) and the high eruption rate of the lunar lavas. The high eruption rate will make the lava flow for greater distances (helped by the low viscosity) and the channels much larger. Low viscosity will also ensure that overflows from the channel spread out rapidly away from the channel itself, so that they do not build up any noticeable levees on the banks. Greater fluidity of the lava will also presumably mean that the channels within it are more free to move about, giving the sinuous rilles their higher sinuosity. The best known of the sinuous rilles is Hadley rille (below right) described by Greeley, which was visited by the Apollo 15 astronauts (above). It is a fairly typical sinuous rille, starting in an elongate crater near the bottom of the picture, diverted by the high ground near the Apollo landing site (marked with a star), narrowing almost to nothing before it again hugs the higher ground of the Apennine Mountains and then fades to invisibility in the same area as another sinuous rille from the opposite direction. The astronauts were able to get a good look at it, and the view revealed a steep scree slope on either side, down which boulders have rolled and collected in the bottom (below left). The sudden narrowing of the rille (arrowed) indicates that, for part of its length at least, the rille was a roofed-over tube, and this narrower section is probably a point where the collapse of the roof was only partial.

Among the pieces of equipment that the Apollo 15 astronauts deployed on the lunar surface was a seismometer. This was the third working seismometer on the Moon, and it meant that natural moonquakes could be pinpointed with accuracy and that the interior structure of the Moon could be investigated. Shocks from the deliberately crashed Saturn IVB rockets on several Apollo missions, together with natural impacts on the Moon, including a very large one of about one tonne on the farside of the Moon in July 1972, have enabled a good picture of the inside of the Moon to be built up. 25km down, there is a sudden increase in the velocity of seismic waves, interpreted as an absence of any cracks below this level — perhaps because they have been annealed. At 60km depth there is another increase in velocity, interpreted as a change in composition of the rocks from anorthositic gabbros above to a pyroxene-olivine layer below. This effectively marks the bottom of the lunar crust. Below this, down to about 1,000km depth, is a very rigid, stable zone, through which seismic waves travel easily, losing very little of their intensity. This solid part of the Moon is known as the lithosphere, by analogy with a similar, much thinner, layer on the Earth. It is at the base of this zone that the moonquakes occur. From about 1,000km depth to the centre of the Moon at 1,738km depth is a partially molten zone analogous to the Earth's asthenosphere.

The Moon: The Orientale Basin

One of the most extraordinary and exciting pictures to be received from the Orbiter probes was that of the 900km-diameter Orientale basin (opposite) on the western limb of the Moon. Basins, such as Imbrium on the visible face of the Moon, had been well known for a long time, but none was as clear and fresh as Orientale, with the beautiful symmetry of its circular, concentric mountain-rings. The origin of these basins, like the origin of the craters, had been much in dispute, some investigators favouring an impact origin but many arguing that they were huge, fault-controlled tectonic structures, modified by volcanic action. This one picture of the Orientale basin, however, did more than anything else to swing the general opinion round to the view that the basins were of the same origin as the large craters, so that, as the craters became more generally accepted to be of impact origin, so did the basins. The rim of Orientale is circular, and there is a clear blanket of ejecta round it, part of which is seen in close-up middle left overleaf, obscuring the older craters underneath and becoming radially striated further out. The resemblance to the ejecta blankets of impact craters such as Timocharis, for example (page 21), is clear. In close-up (insert lower right) finer striations are seen, and also traces of flows resembling those on the crater Tycho (page 29). Outside the ejecta blanket, secondary craters become evident (top right), similar to those of Copernicus (page 23) in their irregularity and in the way they form in chains approximately radial to the basin centre. Note also the way the secondary craters further from Orientale overlap the adjacent nearer craters. The long trough at the bottom of this insert bears a strong likeness to the feature in the top right picture of page 23. Both are chains of secondary craters, with the craters in the chain so close together that they have merged, but in the picture opposite the crater chain is 15km wide as opposed to 3km wide in the case of the Copernicus chain.

These similarities to craters are found outside the basin's rim, but inside it the picture is very different. Instead of the terraces, flat floor and central peak of large craters the Orientale basin has two inner rings of mountains. These have been interpreted in two ways. Either they represent the different layers of the crust that the impact has penetrated, or else they are a combination of concentric fractures and an enlarged central peak. Other large craters or small basins show that, above a critical size, the central peak enlarges to a ring.

Between the two outermost rings of Orientale is a flatter area resembling the floor of an impact crater such as Aristarchus (page 27). The middle of the Orientale basin, like the other circular *mare* basins on the nearside of the Moon, is flooded with dark *mare* lavas. An enlargement of the southern border of this *mare* is shown in the bottom left insert. *Mare* lavas can be seen embaying Orientale material, and curiously shaped hollows with steep stepped edges suggest that lakes of lava were ponded in them for a while, draining in stages to leave the steps at the edge of the lake where the lava solidified.

The Moon: The Imbrium Basin

The huge ring of mountains in the upper half of the picture opposite, comprising the lunar Apennines to the right, the Alps above and the Carpathians below, is nearly 1,300km across, and marks the outer rim of the Imbrium basin, the largest on the Moon. From its more eroded appearance, Imbrium is clearly older than Orientale, and in addition has been much more heavily flooded with lavas, which have obscured many parts of the ejecta blanket and secondary craters and much of the interior also. Nevertheless, they are still visible in some regions — the lower right insert shows an oblique view of the area around the crater Ukert, in the ejecta blanket of Imbrium near the rim. Small-scale striations like those in the Orientale ejecta blanket and Timocharis' ejecta blanket (page 21) are clearly visible. Further out from the crater edge (lower left insert) chains of secondary craters are seen, although they are not so young and well defined as those of Orientale. There is a ring of *mare* ridges in the lavas that fill the interior of the rim, indicating the presence of an inner ring of mountains like that in Mare Orientale, but submerged beneath the *mare* surface. Traces of it show through in the form of isolated mountain peaks or small chains of mountains (arrowed).

Calculations have shown that an impact which produces a crater 1,300km across on a body the size of the Moon (3,476km diameter) is almost enough to break it apart altogether, so the formation of the Imbrium basin must have had far-reaching effects on the surface and interior of the Moon. These effects are seen on the surface as a radial and concentric pattern of fractures that extends over most of the Moon. Nearly one third of all lunar rilles are either concentric or radial to the Mare Imbrium. The effects of the Imbrium impact must have been equally devastating beneath the lunar surface, for the initial crater formed in the excavation stage (page 20) was probably a few hundred kilometres deep before the floor rebounded back again and the inner rings formed. The Moon must have been fractured for a considerable depth, and any later volcanic activity would have used these fractures as routes for the magma to reach the surface and erupt as lava. Sinuous rilles and other volcanic vents tend to cluster near the edge of the circular *maria*, suggesting that most of the lava that filled the basin was erupted from the edge. The large flows with prominent flow-fronts (page 43) that flow into the Mare Imbrium have also been erupted from the edge.

There is also much evidence that the *mare* basins filled with lava gradually over a long period of time. In the picture opposite, large craters such as Archimedes (A) and Plato (P) were formed after the Imbrium impact because they lie on top of it, yet are older than the Imbrium lava fill because they themselves have been filled with lava, indicating that between impact and final flooding there was time for several large craters to be formed. There is also direct evidence from the samples: the Imbrium impact is dated at 3,900 million years, whereas the Apollo 15 samples indicate that Imbrium lava flooding was continuing to occur 3,300 million years ago.

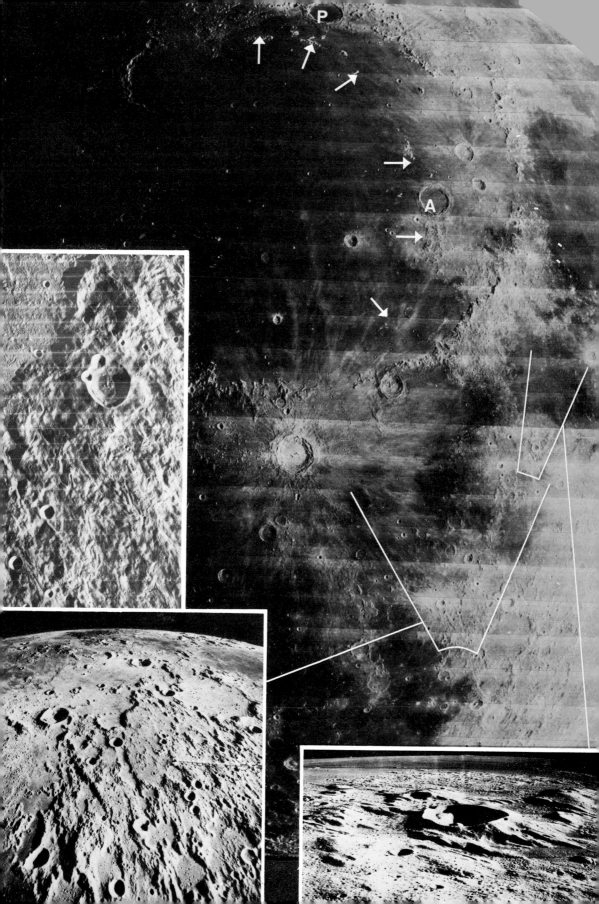

The Moon: Mascons

As the lunar Orbiter spacecraft were orbiting the Moon, precise measurements were made of their positions as they crossed the nearside. In 1968, two Americans, Müller and Sjögren, analysed these data to look for local changes in gravity. They were interested particularly in the larger basins on the nearside, for since these were depressions in the ground they should be the centres of large negative anomalies — that is, the measurements of gravity should give much lower values than the average. To their amazement, exactly the opposite result emerged. Each time the Orbiters approached one of the circular *maria*, they accelerated, and as they moved away had slowed down, showing that the basins were *positive* gravity anomalies. This indicated that there was a concentration of mass rather than a deficiency of mass in the nearside basins, and these concentrations of mass were named mascons. They were the centre of a good deal of interest and excitement, and there was much speculation as to what the concentrations of mass were and why they should be there. Some believed they were the dense bodies of impacting masses buried in the Moon, others that they were vast buried volcanic complexes of dykes, sills and intrusions. As more detailed observations from recent lunar craft have been made, it has become possible to model the shape of the mascons as flat, disk-shaped objects on the surface. The general consensus now is that these mascons *are* the lavas that fill the basins. The fact that the mascons are still there after more than 3,000 million years indicates that the lithosphere of the Moon is thick, having the strength to support the weight of the mascons over such a long period. This rigidity has since been confirmed by the Apollo seismic experiments (page 50). The lunar highlands, on the other hand, show no positive anomalies, so they are in isostatic equilibrium. At the time when the highlands were formed, then, the lithosphere must have been much thinner. Even during the early lava flooding of the basins there was a certain amount of adjustment taking place. Evidence of this can be seen in the pictures opposite. The main picture shows part of the magnificent series of parallel graben round the Mare Humorum. They indicate tensional movement towards the *mare* centre, which could be explained by the sinking of the centre stretching the surface around it. More direct evidence is found in the insert on the left, which shows the opposite edge of the same *mare*. Here a prominent fault bounds the *mare* surface, with the *mare* centre sinking relative to the terrain on the left. Note that an old crater has been cut in two by the fault, and the right side buried beneath the *mare*, whereas the more recent smaller crater straddling the fault below it shows no signs of disturbance. These observations suggest that in the early stages of lava flooding, at least, the basin centre was sinking under isostasy.

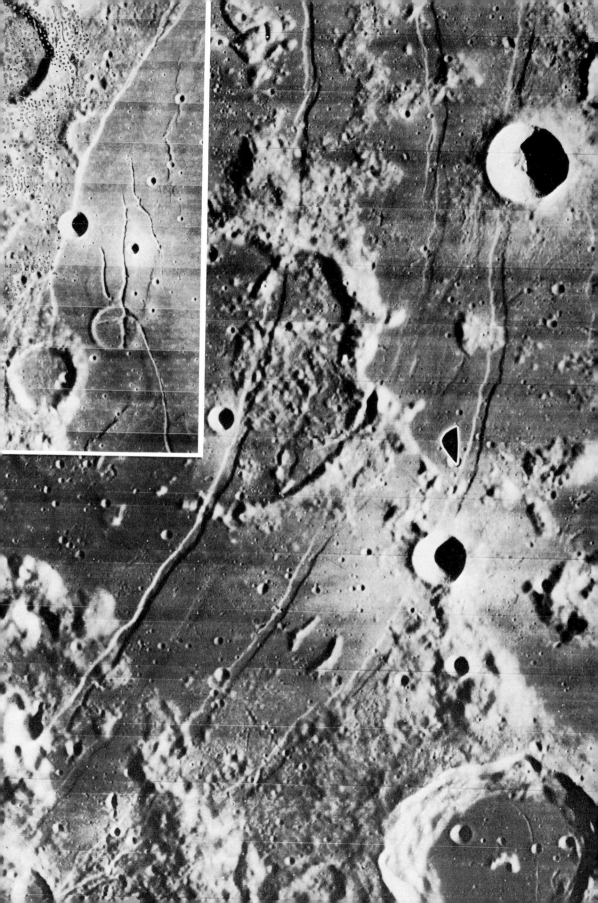

The Moon: Farside Lunar Basins

When Luna 3 took the first ever pictures of the other side of the Moon in October 1959 it was the first spectacular feat in the modern exploration of the Moon. The quality of the pictures seems poor by comparison with later missions, and it was difficult to make out the details (a white marking optimistically named the 'Sovietsky Mountains' turned out to be a crater ray), but one immediately apparent peculiarity which marked the farside of the Moon as being fundamentally different from the nearside was the almost complete absence of the familiar dark areas of the *maria*. There are, however, large circular basins on the farside, though none as big as Imbrium or Orientale, and two of these — Mendeleev (above), 340km across with traces of an inner ring; and Gagarin (below), 260km across — are shown opposite, but both are without the dark *mare* filling of the nearside basins. Why, then, has almost all the volcanic activity occurred on the Earth-facing side of the Moon? Clues to help answer this question were found in other lop-sided characteristics of the Moon. One of the instruments on board the Apollo craft was a very accurate altimeter which, using a laser beam, was able to measure heights to better than one metre's accuracy. The results showed that the Moon is very slightly egg-shaped, with the longest diameter about 2km longer than the shortest. The longest diameter points towards the Earth. It is also found that the centre of mass of the Moon is not at the Moon's centre of figure but displaced 2km towards the Earth. This implies that the crust of the Moon is thinner on the Earth-facing side, which alone may account for the larger area of *maria*. The reason for the lop-sided shape and mass distribution may well be that the Moon was once much closer to the Earth, the greater tidal forces causing this asymmetry. At any rate it has resulted in the Moon keeping the same face permanently towards the Earth, for the Earth's gravity attracts the bulge more strongly than the rest of the Moon.

There has been much speculation as to how the magma was formed that erupted onto the surface to form the *mare* lavas. On the Earth, the lithosphere is thin and formed of separate plates that are slowly moving relative to one another. Along mid-oceanic ridges, either side of which plates are moving apart, volcanoes tend to occur and these eruptions of lava create new crust which then spreads across the Earth's surface. Volcanoes may also erupt where one plate is moving towards another and buckling under it; the frictional heat beneath the surface melts the descending rock to form magma that will eventually erupt as lava. On the Moon, as we have seen, the lithosphere is very thick and stable, so plate tectonics cannot operate at the present, and there are none of the tell-tale signs of it having operated in the past. It is probable that the progressive decay of radioactive elements provided the source of heat for a temperature build-up to the melting point of rocks within the Moon. This melt then erupted onto the surface through weaknesses in the crust, such as the major fractures associated with basin-forming impacts.

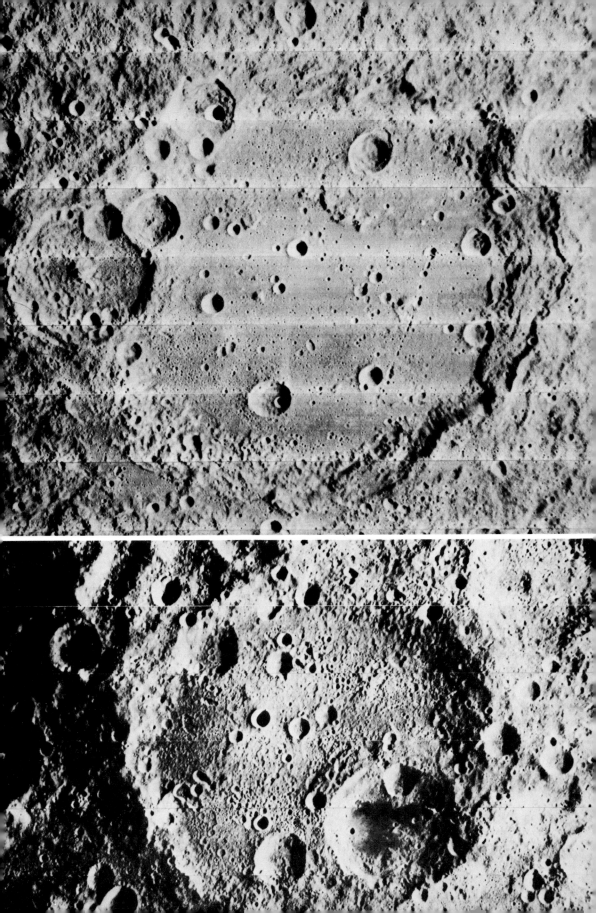

The Moon: The Highlands

Most of the lunar surface is covered with large impact craters, so densely packed that they abut and overlap one another to form a continuously cratered surface known as the highlands. This indicates that a heavy bombardment of large bodies has taken place on these areas, the later craters steadily obscuring the earlier ones until they eventually lose their identity and become no longer recognizable as craters. This effect can be seen in the picture opposite, which shows a highland surface on the farside of the Moon. All stages of crater degradation can be seen, ranging from fresh craters with sharply defined rims: through craters with smoothed-over rims and many subsequent craters obscuring them; to the oldest craters where only a few segments of the rim remain visible through the maze of more recent impacts. It was observed on page 16 that the highlands are older than the *maria*, but the lunar samples showed that the difference in age between the two is not very great — the oldest *maria* are about 3,900 million years and the age of the Moon is 4,600 million years, so the highlands were exposed to bombardment for an extra 700 million years, less than 20% of the total time available for cratering on the *maria*. Yet a brief glance at the picture opposite and the *mare* on page 39 shows that the highlands have far more than 20% more large craters than the *maria* — the figure is several hundred per cent — so it seems that the Moon must have undergone a far heavier bombardment before the formation of the *maria* than afterwards.

To find out why, we must look to the various theories of how the Moon was formed. There are several theories, falling into three basic groups: (a) that the Moon and the Earth formed together as a double planet; (b) that the Moon split off from the Earth; and (c) that the Earth's gravitational pull 'captured' the Moon into orbit from some other part of the Solar System.

There is no general consensus at the moment favouring any one of these theories, for there is little concrete evidence left to help decipher an event that happened 4,600 million years ago. Most theories, however, indicate that the Moon at some stage was bombarded by smaller bodies, and it is probable that the heavy cratering in the highlands is a result of the last stages of this bombardment. It is clear that, when the Moon has orbited the Sun 700 million times, it will have collided with most of the bodies in the Solar System that have orbits passing close enough to hit it. The rate of impacts should, therefore, have declined very rapidly early in the Moon's history, but should be declining very slowly at the present time.

The inset shows a curious level area in the highlands round the crater Cayley. Areas like this which are bright but smooth are known as Cayley plains and are found in many other places in the highlands. Their origin is discussed later (page 64).

The Moon: Dating the Highlands

Despite the great age of the highlands, clues as to what the early structure of the Moon was like have been systematically destroyed by the intense bombardment of its surface. This has smashed up the rock, thrown it great distances across the lunar surface, completely altered its structure and appearance, heated it and, in some cases, melted it. The process of altering a rock suddenly by transient high pressures and temperatures is known as shock metamorphism. All the rocks recovered from the highlands have been shock-metamorphosed breccias. An important consequence of melting and shock metamorphism is that it will alter the measured age of a rock. Rocks are dated by measuring the amount of decay of radioactive elements: a cataclysmic impact event will mean that many of these 'age clocks' are reset, so that the rock will on analysis give the age of the impact event rather than the age of the rock's origin. Thus many highland rocks yield ages of around 3,900 million years—close to the date of the Imbrium impact—but there are a few older rocks with ages as old as 4,200 million years, and just one sample gives a date of 4,600 million years—the age of the Moon itself.

There is much evidence for volcanic activity in the highlands, but most of it is found in the regions bordering the *maria*, and appears to be associated with marial volcanism. Good examples of this type of activity are the crack-floored craters, such as the crater Alphonsus, 115km across (above opposite), which tend to concentrate round the edges of the marial regions. From the morphology of crack-floored craters it is clear that they are of impact origin, and often they are old and eroded like Alphonsus, but the width and depth of the cracks on their floors distinguishes them from ordinary impact craters. In some of them, the floor has been raised to form a plateau within the crater walls, and it is quite common to find the crater partially flooded with dark *mare* material. All these characteristics can be explained by intrusion or eruption of lava, the fissures beneath the impact crater being used as dykes for the magma in the same way as they were for the large *mare* basins. Some of the magma erupted as lava, but that which did not formed sills and other intrusions beneath the crater floor, pushing up the level of the crater floor and forming the fissures that characterize the crack-floored craters.

In Alphonsus, an additional type of volcanism is present in the form of small but prominent dark areas with ill-defined edges, centred around small craters, mostly near the edge of Alphonsus' floor (above). A close-up of two of these dark haloed craters, which each measure about 2km across, is shown below. It is clear that these craters are volcanic: they all occur on fissures, they are not circular in shape, and they are surrounded by dark material that is not formed into rays. There is no sign that lava was erupted from these craters, indicating that they are purely pyroclastic volcanoes; that is, that they were formed by explosive eruptions of material.

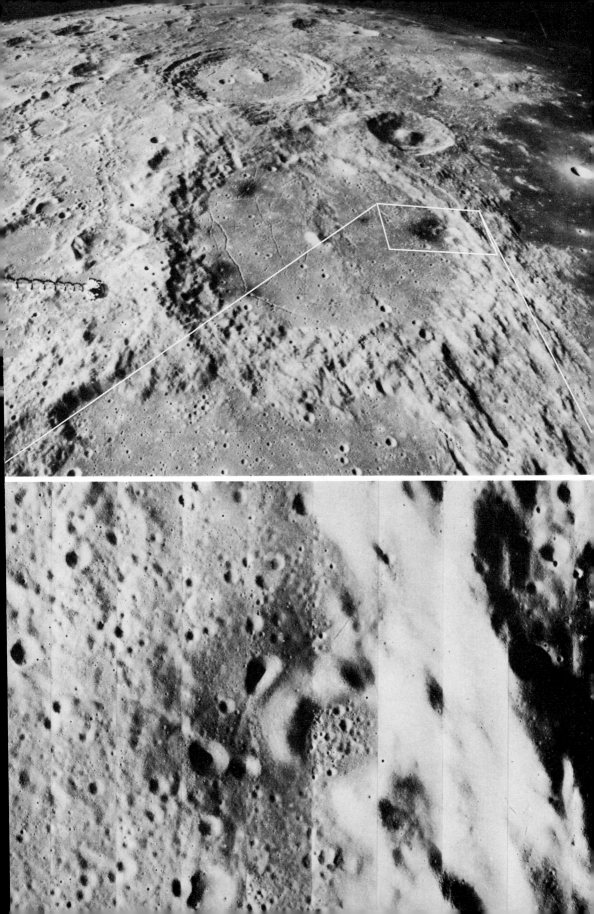

The Moon: Highland Sinuous Rilles

Other volcanic features found in highland regions are sinuous rilles which, like crack-floored craters, are usually found close to the margins of the *maria* and are probably associated with marial volcanism. Two examples are shown opposite: the upper one is close to the Mare Frigoris and the lower one flows into the Mare Imbrium near the crater Plato. In both cases the picture shows an area about 110km across. These rilles are puzzling, for they run through the highlands, which are not, as we have seen, composed of lava, so the rilles cannot be lava channels or tubes as described on page 48. They are, however, so like the sinuous rilles in the *maria* that an origin greatly different would be improbable. It has been suggested, therefore, that these rilles within the highlands were formed as a result of erosion, not by flowing water as river channels are, but by flowing lava. Low viscosity and high eruption rates would mean that a lava erupting in the highlands would flow very rapidly downhill, taking the path of least resistance between topographic obstacles until it reached one of the *maria*, or a topographic depression where it would pond to form a lava lake. If the eruption was prolonged, and the temperature high enough, the stream of hot lava would begin to melt its bed, and start down-cutting into the highland rock over which it was travelling. This may have happened with the two sinuous rilles opposite.

Other signs of volcanism in the highlands were thought to be the Cayley plains illustrated on page 61. They were interpreted by some workers as ash flows, and interest in this possible widespread highland volcanism was so great that an area of Cayley plains was chosen as the landing site for the Apollo 16 mission. None of the rocks brought back from this mission were of primary volcanic origin, however, which ruled out the possibility of ash flows. The origin of these smooth highland plains is still something of a mystery. Some propose that they are smooth ejecta from the Orientale impact basin, but it is probable that there is too much Cayley plains material distributed around the Moon to be explained by one impact event. Others have suggested that secondary cratering may be responsible, for secondary impacts striking hills will start landslides of material into the valleys, filling them up with smooth deposits. The mystery is enhanced by a similar widespread deposit that is found on the planet Mercury.

To sum up, the highlands are made up of igneous rocks, modified by impact, and they cover most of the Moon's surface. The major geological provinces of the highlands, indeed of the whole Moon, are those of the big basins. Impact is the process which dominates the Moon's present surface and past history.

Mercury: First Encounter

Mercury is the nearest known planet to the Sun, orbiting at a mean distance of 57.9 million km, and is never more than 28 angular degrees from the Sun. Consequently it is difficult to study using Earth-based telescopes, and thus is the least studied of the terrestrial planets, although its existence has been known for thousands of years. However, some important observations have been made from the Earth. Telescopic observations showed, for example, that Mercury is small, with a diameter of 4,880km, about one third that of the Earth, but has a high density, of 5.45g/cm³. This is much higher than that of the Moon or Mars, and closer to that of the Earth, which implies that a similarly high proportion of the planet is made up of iron.

However, earthbound observations do not indicate whether the iron is distributed homogenously with a rocky silicate phase or if the planet is differentiated into an iron core and rocky mantle as is the case with the Earth. Certainly the upper few centimetres or metres, at least, have the characteristics of silicate rocks like those on the Moon, broken up by meteoritic impact to produce a regolith. This is fairly certain because the reflection of visible and radio waves from the Sun off the mercurian surface and the thermal emission of the surface at infrared and radio wavelengths are similar to those that would be expected of the Moon if it were placed in the same orbit as Mercury. These polarimetric and photometric simi-

larities also imply a lack of atmosphere around the planet. The degree of meteoritic bombardment which would have caused the inferred regolith could not be determined telescopically as the resolution so obtained is at best about 300km, and all that can be seen are faint light and dark markings (see page 71). Studying the rate of movement of these markings enabled astronomers to calculate that Mercury rotates exactly three times while circling the Sun twice, giving it a day which is two of its years long. The mercurian year lasts 87.9 Earth days.

It was seen from Earth that one hemisphere of Mercury tends to be more 'contrasty' than the other, and Mariner 10 observed a part of both the apparently different faces. This spacecraft was launched on 3 November 1973. Immediately after launch it photographed both Earth and Moon as it sped on its way to swing around Venus (which it photographed in passing) and in towards Mercury. The first encounter with the planet in March 1974 was an equatorial pass, 700km at closest approach, which occurred on the dark side. The photomosaic opposite is composed of pictures taken on this first visit, just before Mariner 10 flew past the dark side. This view, of part of the 'blander' hemisphere of the planet as seen from Earth, and that on the next page of the more 'contrasty' hemisphere, showed that the surface of Mercury is Moon-like, the dominant landform being craters.

Mercury: The Surface

As the spacecraft came back into the sunlight on the other side of the planet, it took more photographs, some of which were put together to make up the photomosaic shown opposite. The craters range in size from the limit of photographic resolution (100m at best) up to the largest basin, Caloris, which is 1,300km across. The half of Caloris sunlit during the encounter can be seen opposite, the other half being hidden beyond the terminator (the boundary between the light and dark hemispheres of the planet).

Surrounding the basin is an aureole of relatively smooth terrain. Geologists such as Robert Strom and Newell Trask on the Mariner 10 mission considered this a distinct geological unit, which they called Smooth Plains. This unit is thought to have been produced by the eruption of large amounts of lava, occurring after most craters, including Caloris, had been formed.

Beyond the swathe of Smooth Plains is the heavily cratered terrain. The craters are excavated in an early mercurian surface which can be seen in between the craters, and is hence called the Intercrater Plains.

Observing these features and their similarities to the Moon, the Mariner 10 science team concluded that planetary differentation had occurred early in the history of Mercury, giving rise to a silicate crust and nickel-iron core. The strongest evidence for this was one of Mariner 10's most unexpected findings: that Mercury has an intrinsic magnetic field. This was revealed by plasma and charged-particles experiments and the magnetometer. It is not clear, however, whether the field is produced by a planetary process occurring now or whether it is a relic of an ancient field.

The lack of any atmosphere, suspected from Earth-based observations, was confirmed during the occultation of the Mariner 10 spacecraft as it flew behind Mercury during the first encounter. Radio signals from the spacecraft were blocked by Mercury in the manner expected for a planet with no atmosphere. The lack of atmosphere was confirmed by ultraviolet spectrometer observations which did, however, detect minute quantities of helium. No landforms suggesting atmospheric erosion were observed and it was concluded that any atmosphere Mercury might have had was lost very early in its history. Occultation also enabled the scientists on the celestial mechanics experiment to show that Mercury is much closer to being a perfect sphere than are the Earth or Moon. This sphericity adds weight to the speculation that the surface features seen on the sunlit hemisphere are representative of the general appearance of the planet; for, as we have seen, the very different Earth-facing hemisphere of the Moon is accompanied by a prominent bulge in the Moon's figure.

Caloris

Mercury: The Southern Hemisphere

The second encounter of Mariner 10 with Mercury occurred six months after the first, in September 1974, and was arranged to pass over the south pole at about 50,000km. Pictures taken on this pass form the mosaic opposite. Because Mercury had orbited the Sun twice in the time it had taken the spacecraft to orbit it once, the same hemisphere of the planet was illuminated by the Sun, as was the case for the third and last pass over the northern hemisphere at closest approach of only 370km (March 1975). This had the advantage of different camera viewpoints on the three passes, ensuring good coverage of almost all of the one hemisphere. Furthermore, by comparing the lengths of shadows of the same features photographed at high resolution in more than one of the encounters, it was shown that earlier estimations of Mercury's rotation rate were correct.

It is the Mariner 10 photographic record that forms the basis of this chapter. Since even the most sharp-eyed observers with the best telescopes could only produce maps with a few vague smudges, albeit with beautiful classical names, it is clear from the photography in the following pages exactly how large an advance Mariner 10 represents. One of the last of these maps was done by John Murray in 1972 and is shown below, opposite. The large dark circular feature in the centre is on the more 'contrasty' half of Mercury and corresponds to the Caloris basin shown on the previous page. (The map is upside down relative to the picture, however, owing to the convention then followed by astronomers.)

The imaging system itself produced too much information to be represented on one picture, and so specialized processing techniques and procedures were used to display the particular information desired. Some of these are discussed later. Areas overlapped in first and second encounter pictures provided data which could be processed into stereo pictures.

An unusual discovery from Mariner 10 photography was the presence of long, sinuous scarps, which are interpreted as the surface manifestations of low-angle thrust-faults. One of the largest, Discovery Rupes (named after Captain Scott's ship), can be seen clearly (arrowed) on the mosaic opposite.

In this view of the southern hemisphere of the planet, bright rays can be seen radiating from many craters. These craters comprise the youngest geological unit on Mercury and are the only substantial events to have occurred on the planet in the billions of years since the formation of the Smooth Plains.

One thing the imaging scientists, led by Bruce Murray, looked for but failed to find was a mercurian moon, and it was concluded that if any orbiting body exists it must be less than 5km in diameter.

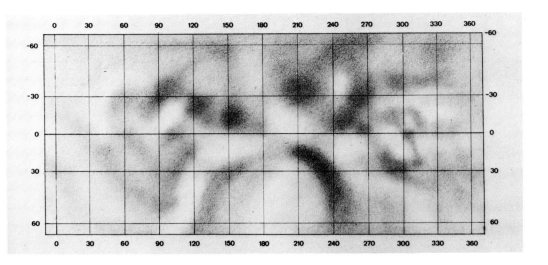

Mercury: Small Impact Craters

The discovery, not unexpected, that the surface of Mercury is pock-marked with craters confirmed that the impact process has been an important one in the geological histories of solid planetary bodies in the Solar System. Mercurian craters have all the morphological elements of their lunar counterparts. One characteristic of very fresh craters on both the Moon and Mercury is the bright ray system which radiates from them; for example, that shown in the picture opposite, top right. Mercurian craters also exhibit a similar progressive change in morphology with size. Thus the smallest craters, from microscopic size up to about the 8km diameter (opposite, crater A below left) are bowl-shaped. Slightly larger craters such as crater B, below left (about 10km in diameter) may have small flat floors. Still larger craters have central peaks like that in the crater shown below centre. The unusual crater in the photograph below right has a small concentric bench-like ring inside it. This may be caused by the crater excavating different layers of rock which have varying degrees of strength; in this case the crater is within Smooth Plains and the bench may mark the point at which the crater has excavated through the overlying regolith to a hard volcanic layer.

In the larger craters, along with the peaks, terraces begin to form (see page 26).

These are just about visible in the crater (b), shown at top left, which is about 20km in diameter. A hummocky ejecta blanket is present around this crater, and just visible further out are tiny secondary impact craters. The fact that such small craters have not yet been eroded away by subsequent meteoritic impacts is a good indication that the crater is comparatively young. This picture illustrates the ageing progression of the craters, which is similar to that on the Moon. Thus the most recent craters are the small fresh bowl-shaped craters like (a). The next oldest crater in the area covered by this photograph is (b), the ejecta blanket of which has been pocked by crater (a). Crater (c) is more subdued in appearance than (b), has no visible secondaries although it is about the same size, and is thus older still. The terraces within (c) are beginning to lose their individual distinctness, and are coalescing. This process is virtually complete in the very old crater (d), while the oldest crater (e) is marked only by a featureless, roughly circular depression. By analogy with the Moon, therefore, crater (e) must be more than four billion years old.

The small crater (a) is called Hun Kal, which is the Mayan for twenty (the Mayans used a base-twenty number system), because the twentieth meridian is defined as passing through its centre.

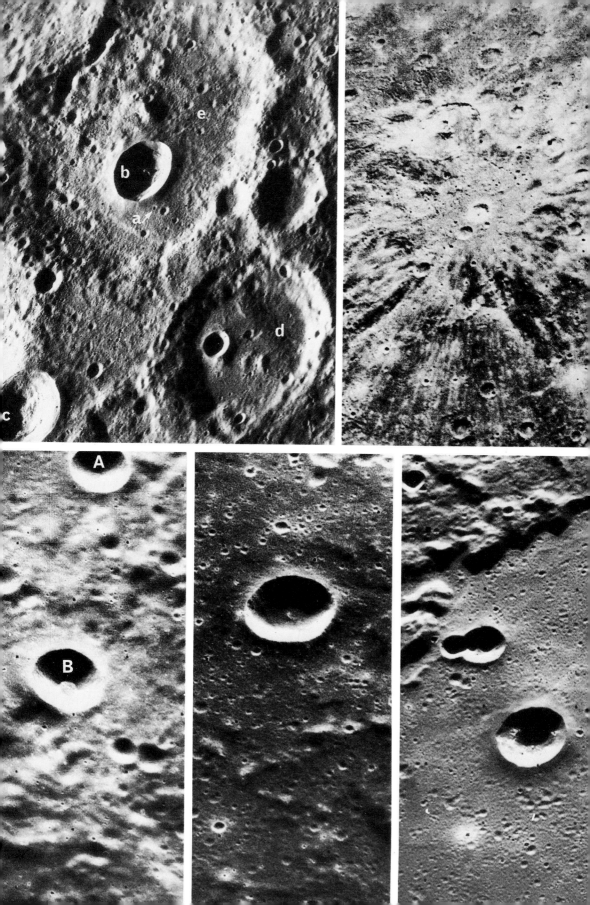

Mercury: Large Impact Craters

The large crater Brahms, pictured at top, is about 95km across and exhibits all the well documented characteristics of a 'fresh' impact. The secondaries are more clearly defined than for the smaller crater (b) on the previous page, both because they are larger and because they have been excavated from a smoother flatter surface, the Smooth Plains (see page 68). The secondaries can be seen quite close to the rim of the crater. This is characteristic of mercurian craters, and is suggested by geologists like Don Gault to be a result of the high gravity, which tends to restrict the ejecta trajectories so that the hummocky ejecta is closer to the crater than would be the case on the Moon. This has two consequences: firstly, the secondaries will have a larger erosive effect than their lunar equivalents, because they are more densely spaced; secondly, the hummocky ejecta blankets will be less likely to overlap each other and obscure the terrain underneath.

Inside the crater is a heavily terraced crater wall which gives way to a hummocky floor. This hummocky material is similar to that found in many lunar craters and is interpreted as a mixture of impact-melted and -brecciated target rocks which has collected in the bottom of the crater during and after excavation.

The twin peak in the centre of the structure is an example of the more complex peak morphologies found in larger craters. Another example can be seen at lower right. This crater has a half ring enclosing a peak. There are several examples of this type of prominence in craters of about 100km diameter on Mercury, but few on the Moon.

Some large mercurian impact structures can show more unusual features, just as with the smaller craters discussed on the previous page. For example, many have a dark halo and dark and light interior markings, like the two in the lower left picture — 100km-diameter craters Balzac (top) and Tyagaraja (bottom). Scientists such as Bruce Hapke, specializing in colour observations on the Mariner 10 mission noted that the dark areas tend to be redder than the surrounding region, while the light patches tend to be bluer. The redder materials have been interpreted as being compositionally different from the surrounding terrain either due to a different layer (presumably volcanic) being excavated by the crater or, in the case of the interior red materials, possibly due to later volcanic infill. The bright, bluer patches have been interpreted either as intensely shock-metamorphosed crustal materials or as regions which have been chemically altered by fumarolic activity — that is, hot volcanic gases percolating through the surface. Unusually coloured craters elsewhere on Mercury show additional features, such as upraised floors, which, in their lunar counterparts, are indications of volcanic activity.

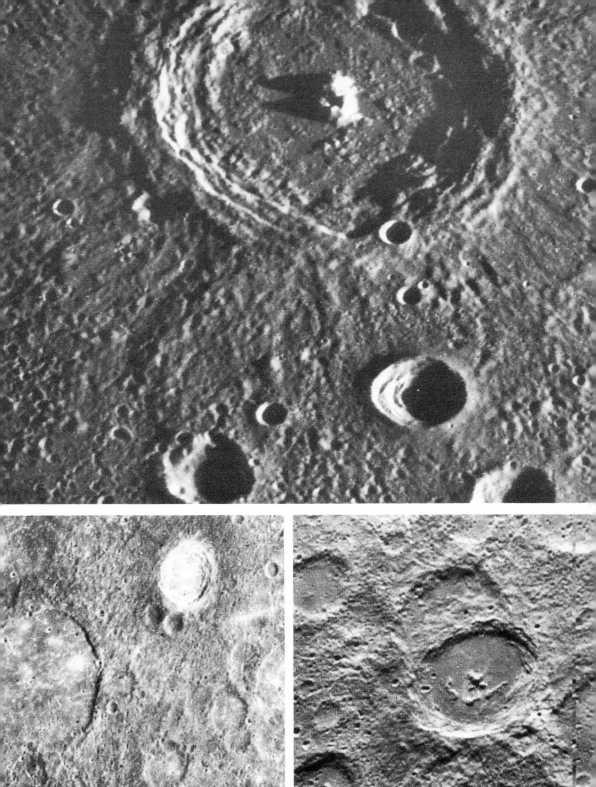

Mercury: Atypical Impact Craters

There are several types of crater on Mercury which are associated with scarps. The upper right photograph shows a crater about 35km across, which has been excavated from the rim of a much larger 100km crater. Roughly concentric to the smaller crater (and to our left of it) is an outward-facing scarp several hundred metres high, bounding a lobe of rugged terrain which occupies part of the smoother floor of the larger crater. The lobe has striations within it which are concentric to the rim of the smaller crater and it appears to be a debris slide formed when the force of the impact of the smaller crater pushed the brecciated material in the rim of the older crater down onto its floor. Large debris slides are well known on Mars, where they are thought to be aided by the presence of an atmosphere and perhaps permafrost. Almost certainly such conditions did not prevail on Mercury at the time these features were formed, which may explain why such features are rare on Mercury. Another possible example is shown in the upper left photograph (arrowed), although the irregularity of this scarp and the lack of concentric striations behind it indicate that it may be a fault scarp which has by chance formed in that position.

The picture at lower left shows another type of scarp (arrowed) associated with some craters. In this case the inward-facing scarp is very small (perhaps 100m high) and roughly concentric to the rim, occurring about half way in towards the central peak. The scarp separates Smooth Plains in the centre from hummocky material nearer the walls. It might be speculated that magma came up through fractures induced in the crust by the impact, and was erupted as lava from vents on the crater floor, leaving this level plain around the central peak. Once emplaced, the weight of the volcanic material may have pushed down the floor of the crater slightly leaving a small scarp at the edge of the Smooth Plains.

Any explanations for the features shown in the lower right picture must also be speculative. In this case several ancient crater rims protrude above the surrounding Smooth Plains, and are bounded by curved scarps (arrowed). The rims are covered in hilly material, probably ejecta from the Caloris impact. Whatever other processes have helped to form these structures, so far found only on Suisei Planitia, 1,000km north of Caloris, it is likely that, when the Smooth Plains were undergoing tectonic modification, fault scarps formed preferentially at the inhomogeneities produced by the crater/Smooth Plains geological boundaries.

Mercury: Linear Crater Walls

The nature of the terrain in which a crater is excavated (the target material) can be important in determining the final shape of the crater. Hitomaro (top) (about 100km diameter) is a good example of this. Here a stress pattern or set of vertical faults present in the target material caused scarps to form in the rim of the subsequent crater.

Central peaks occupy only the western (right) half of Hitomaro. This is possibly an example of asymmetry induced by an inhomogenous target material, since the crater was impacted into the edge of a much larger basin. The western half of Hitomaro was underlain by the basin rim and formed a group of normal central peaks. However, the eastern half, underlain by the basin floor, was at a lower level and did not produce any peaks, or at least none large enough to protrude above the Smooth Plains which later floored Hitomaro. Other plausible examples of induced asymmetry are seen elsewhere on Mercury.

The Smooth Plains within Hitomaro look too smooth to be impact melt-breccia, especially when compared with the hummocky material within Brahms, shown on page 75. It is difficult to envisage them as anything other than volcanic in origin, although the small patches of material perched on the terraces may be impact melt. The plains have a number of irregular rimless pits (r—compare with small crater, c), which on the Moon often indicate volcanic deposits (see page 44). The small scarp (s) along the base of the crater wall is probably the top of a buried terrace.

100km south of Hitomaro is the crater Mahler (M in lower right picture). This 100km crater has a linear wall segment trending east-northeast—the same direction as those on the rim of Hitomaro (the photograph in this case is oriented with north to the right). It also has one tending west-northwest. This 'grid' pattern seems to be present over much of the planet's surface, and there are many other craters with linear wall segments. Some are almost perfect hexagons rather than the usual circular shape expected for impact craters. Clearly such indications of tectonic patterns are very useful in deciphering the geological history of the planet.

The picture at lower left shows two craters which have linear wall segments of a different origin. The shape of the larger crater may be largely determined by the underlying inhomogeneity between old basin rim material (R) and Smooth Plains (S). The smaller crater has an unusual shape which is due to an underlying fault (which has also caused the small scarp (F) running across the Smooth Plains). This is a low-angle thrust fault, and is caused by compression of the planet's crust. It is widely separated from other thrust faults, in contrast to the close spacing of the vertical faults seen in the rim of Hitomaro.

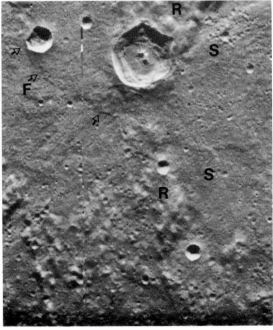

Mercury: Caloris

The largest impact structure on Mercury is the 1,300km basin Caloris, half of which is seen in the photomosaic opposite. The name is Latin for heat and was given to the basin because it is near the subsolar point at perihelion (i.e., when the planet is at its closest to the Sun). The mosaic is made up from all the available Mariner pictures of this region, and some of the pictures composing it show more detail than others because they were taken when the spacecraft was closer to the planet.

It is in some senses unfortunate that, during all three encounters with Mercury, half of Caloris was hidden. However, this had the advantage that the half we do see is under a low angle of illumination, picking out all the fine detail of the basin. Interpretation of the nature of the basin was made much easier by having had the experience of studying big basins on the Moon such as Imbrium and Orientale. There are strong similarities between Caloris and Imbrium, but there are dissimilarities, too, possibly resulting from the much higher surface gravity on Mercury or from differences in the structures of the crusts.

The basin is surrounded by a range of mountains similar to the Apennines around Imbrium (page 55). They rise 2 or 3km above the floor of the basin and, unlike the rims of smaller craters and basins, present a very jagged and lumpy surface, apart from in one region where there is a noticeable gap (G) in the mountains. While some of the mountain blocks seem to be virtually *in situ*, many could be overturned slabs from the edge of the excavated area, while others might be totally dislocated blocks thrown out during the impact. Note the block (B) which seems to have been slightly dislocated from the rim.

The floor of Caloris is unlike anything we have seen so far. Clearly, the basin has been partly filled by materials that formed plains, but these materials were then broken up into concentric and radial ridge patterns on the outer parts and, towards the centre of the basin, networks of huge cracks.

The Caloris impact took place on an already well cratered surface. For example, we see the large crater of Van Eyck (V), the rim of which has clearly been cut by Caloris ejecta, although later volcanic activity has filled the floor of the crater with Smooth Plains. It is likely that Caloris was the last of the big impact basins on Mercury, but it is clear that smaller impacts continued to occur, since craters are superimposed on the rim and floor of the Caloris Basin.

Mercury: Pre-Caloris, Syn-Caloris and Post-Caloris Terrains

The photographs on the opposite page show rock units of three distinct ages. There are features which existed before Caloris was formed, surfaces produced during the Caloris impact, and material emplaced afterwards.

The top picture shows the rim of Caloris — the Caloris Montes. This consists essentially of rock, albeit brecciated and metamorphosed, which existed before the impact. Any pre-existing features have been destroyed during the catastrophic event, of course, and so these mountains are included within the Caloris geological group.

Within the ring of mountains are the plains which fill the basin. Judging by their generally smooth, level appearance and by the relative paucity of superimposed craters they are virtually of the same age as the Smooth Plains outside the basin; for example, those in the picture beneath. All of the craters on the interior plains are superimposed; there are no underlying craters protruding through the surface. This suggests that there was not enough time between the formation of the Caloris basin and the plains within it for any meteoritic impacts to occur. The impact almost certainly excavated deeply enough to bring up material from the hot mantle, but how much this has contributed to the final unit is not clear.

The most interesting features on these plains are the large number of ridges and crevasses which criss-cross the surface. The ridges show some similarities to those on the lunar *maria* and these may likewise have been caused by compression of a surface lava unit. The cracks cut across the ridges and therefore occurred afterwards. They are steep-sided and flat-bottomed, looking like graben structures caused by cooling and contraction of the material in the basin. But the scale is much larger than anything seen on any of the other planets, and it has been suggested that isostatic movement of the plains produced tension which resulted in the polygonal cracking.

The lower picture shows the rim of a basin — Van Eyck — which existed before Caloris but was far enough away to survive the cataclysm. The rim and the surrounding terrain are strongly lineated. The lineations mark great gouges cut by material ejected from Caloris and, possibly, some of the linear ridges are composed of ejecta which was deposited at the same time. Within the basin and in all the low-lying regions surrounding it are Smooth Plains, probably lava extruded very shortly after Caloris.

An interesting feature within the Smooth Plains is an escarpment (arrowed) which faces the rim of Van Eyck and follows it for at least one quarter of the circumference. Another small scarp in the middle of the basin parallels the first one and both seem to have been influenced by underlying topography.

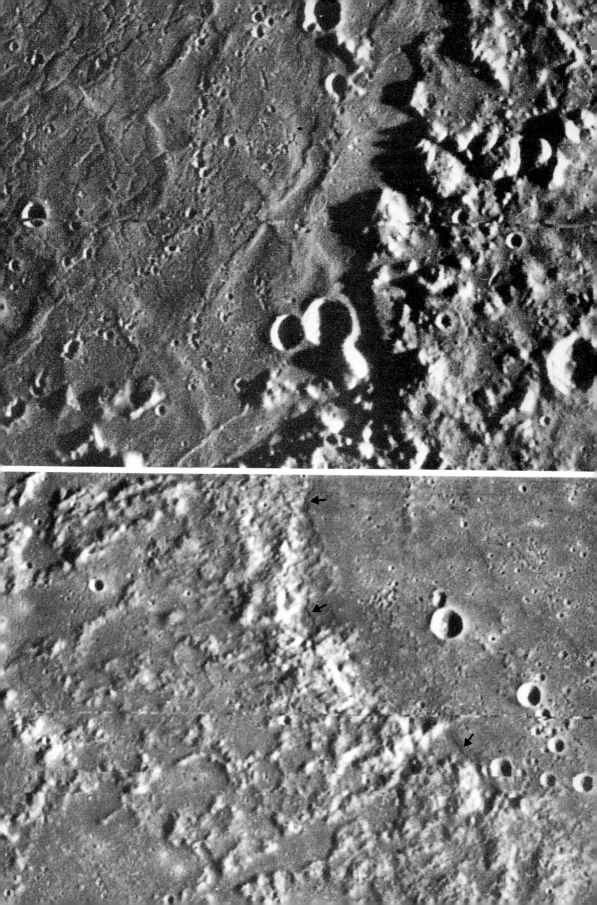

Mercury: Caloris Ejecta

A high-resolution picture of the Smooth Plains can be seen in the upper left photograph. This shows that at this scale they are not so smooth, being pockmarked by many small craters, the smallest resolvable in this picture being about 100m in diameter. Some of these Smooth Plains can be seen embaying a much hillier unit at top right. This material is called, naturally, the Hilly Terrain; underneath it are old pre-Caloris craters whose circular outline is occasionally visible. The boundary between Smooth Plains and Hilly Terrain occurs between 600 and 800km from the rim of Caloris, and the Hilly Terrain is thought to be the outermost of the Caloris ejecta deposits. Between this unit and the basin rim is another Caloris ejecta unit which occupies a large part of the swathe of plains that surround Caloris. This consists of plains with many small hills which are several tens of thousands of metres high and steeper than those of the Hilly Terrain; they are closely spaced in some areas and widely separated in others. This unit, called the Hummocky Plains (centre picture), may contain large lumps of material thrown out by Caloris, forming the hills which seem to protrude from the more level surrounds. The Hummocky Plains surround the outside of the gap in the rim of the basin, and the many small scarps found in the Hummocky Terrain in this region have been suggested as flow fronts formed in the then still semi-molten material as it flowed down to the gap in the basin rim. Like the Smooth Plains with which they are in intimate contact and from which they are sometimes difficult to distinguish, they fill pre-Caloris craters.

How much material was thrown out by Caloris, and what proportion of this was impact melt or impact breccia, is not clear but there must have been vast volumes, some of it thrown right around the planet, and some lost into space. The ejecta deposit nearest to the basin, however, is the material which lies between, and draped upon, the mountains of the basin rim (bottom picture). The smooth surface suggests that this might have been the most molten of the ejecta material, while its undulating surface and topographically high position distinguish it from the otherwise similar Smooth Plains volcanic unit.

(Both the central and lower pictures are long and thin. This is because they were taken on the third encounter with Mercury, during which a fault on Earth-based equipment allowed only a third of the available data to be recorded.)

One Caloris ejecta unit not readily visible comprises the large secondary craters one might expect. Why this is so is not quite clear—it may be that they were covered by Caloris ejecta or the Smooth Plains. However, some probable Caloris secondaries are shown (arrowed, S) at the top right of page 87.

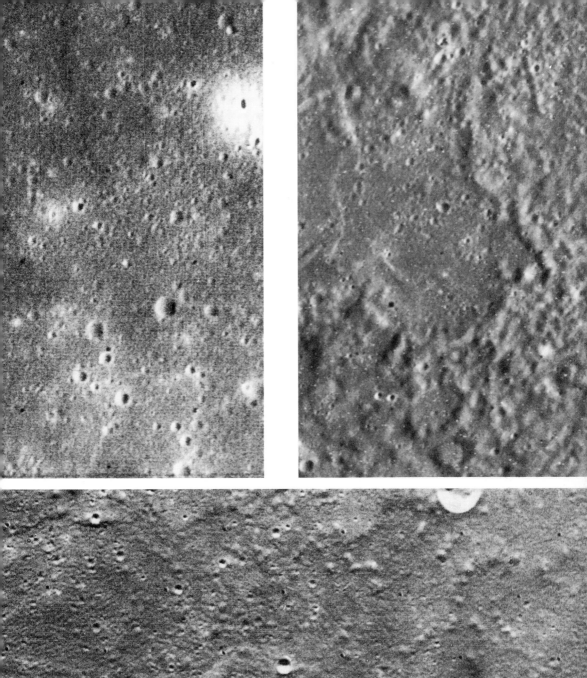
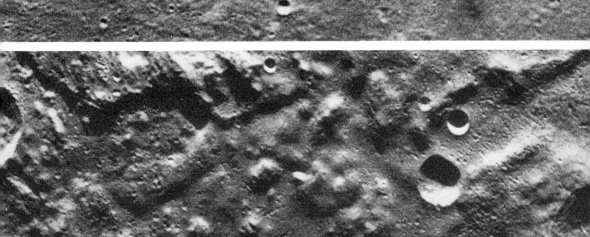

Mercury: Large Basins

The 600km-diameter basin called Beethoven (part seen in upper left) is generally accepted as the second largest basin known on the photographed hemisphere. It appears to be older than Caloris, but it is difficult to make comparisons because the scale is so different, as are the lighting conditions and resolution under which the two structures were viewed. These factors also make it difficult to say how old the plains that fill the basin are. They look fairly level, and smooth enough in some patches to be termed Smooth Plains, but in other places they look rougher. The dark halo around the crater near the middle of the basin may suggest that the crater has excavated a layer of different composition from beneath the surface of the plains. If such layering exists it is generally accepted that this must indicate volcanic activity, in the absence of any other layering agents such as wind or water. The elongate crater (c) within the basin was probably caused by a very oblique meteorite impact. Around the edge of the plains is an outward-facing scarp.

Another of the larger mercurian basins is Tolstoj (440km diameter, upper right). This basin is even older than Beethoven, so old in fact that the rim is scarcely discernible except where ejecta gouges are visible. The centre of the basin is occupied by Smooth Plains (note the rimless pit p), and around this is a dark annulus of Intercrater Plains which extend across the rim.

The dark annulus is bluer in colour than the surrounding region and the Smooth Plains are redder. Such colour variations may be indicative of previous volcanic activity since similar relationships are found in and around some lunar *maria*. Some features (S) north of the basin have been interpreted as secondary chains from Caloris.

Raphael (350km diameter, lower left) is similar in some ways to Beethoven—fairly degraded rim, floored by a plains unit which includes a small crater (c) showing colour variations. The most interesting things about Raphael, however, are the linear features (arrowed) which cut across the floor and rim. These look like the traces of vertical faults, which are unusual on Mercury.

Dostoevskij (390km diameter, lower right) is another large basin filled with plains, although these are not smooth enough to be termed Smooth Plains. As with Beethoven, these plains have an outward-facing scarp running around the rim (arrowed). Such scarps may be indicative of movements in the crust, or perhaps a pushing up of the surface by intrusion of volcanic material below it. Also like Beethoven, radial ejecta gouges are visible. These large basins—unlike smaller ones (see overleaf)—do not display central prominences and this may be due to their size or to the volume of volcanic (or other) fill they contain.

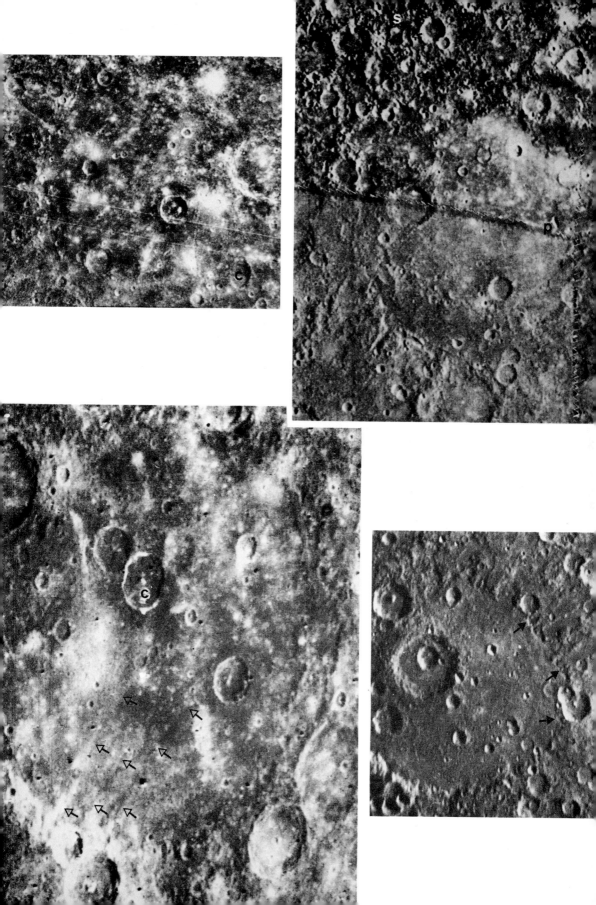

Mercury: Ringed Basins

The transition from crater to basin occurs at about 150km diameter on Mercury, at which size the central peaks of craters are expanded into rings of mountains concentric to the rim. Several of these basins are illustrated in the photographs opposite.

Upper left is Chekhov (200km diameter), a typical example of a multi-ringed basin. No hummocky ejecta blanket or secondary crater chains are visible, and the rim is rather battered. This, and the relatively large number of craters superimposed upon it, indicate that the basin is very old. There are several things of note in Chekhov; for example, there are Smooth Plains inside the inner ring but Intercrater Plains occupy the gap between ring and rim. One part of the wall is linear (L) while the other is affected by a small fault scarp (arrowed).

Top right is an oblique view of two basins, Strindberg (190km), and Ahmad Baba (130km), whose relative youth is evidenced by their generally fresh appearance and well preserved secondary crater chains. Even these contain Smooth Plains, however, indicating that they are a few billion years old.

Centre right is an oblique view of the basin Wang Meng (170km) under a very high lighting angle. This reveals dark patches along part of the inner ring. One scientist, Schultz, has suggested that this resembles the dark mantling around some lunar basins. Such material was sampled at the Apollo 17 landing site, and is considered to originate from pyroclastic and/or fire-fountaining eruptions of *mare* basalt.

Bach (225km, centre left) is a ringed basin with an unusual oblong extension. This probably indicates that it was formed in a very low-angle impact. The secondaries are quite fresh, in contrast to the basin Ma Chih-Yuan (bottom left) which, like Bach, has Smooth Plains at the centre but whose ejecta surface is very degraded, making it more like Chekhov in maturity. Unlike Chekhov, however, the floor is very shallow and appears to have undergone some sort of uplift process possibly caused by intrusion of volcanic materials below.

Schubert (180km diameter, lower right), on the other hand, is clearly quite deep and has a very deep fill of Smooth Plains. These even cover the inner ring though one or two knobs of this protrude through the surface. Note the linear features in the wall (d), which have been interpreted as dykes, and also the crater (c) near Schubert which has dark and light colour markings.

The picture showing Schubert started out as a rectangular photograph of an oblique view of the basin, like that of Wang Meng above. This was ingeniously computer-processed by Mission scientists to produce an orthographic or overhead view.

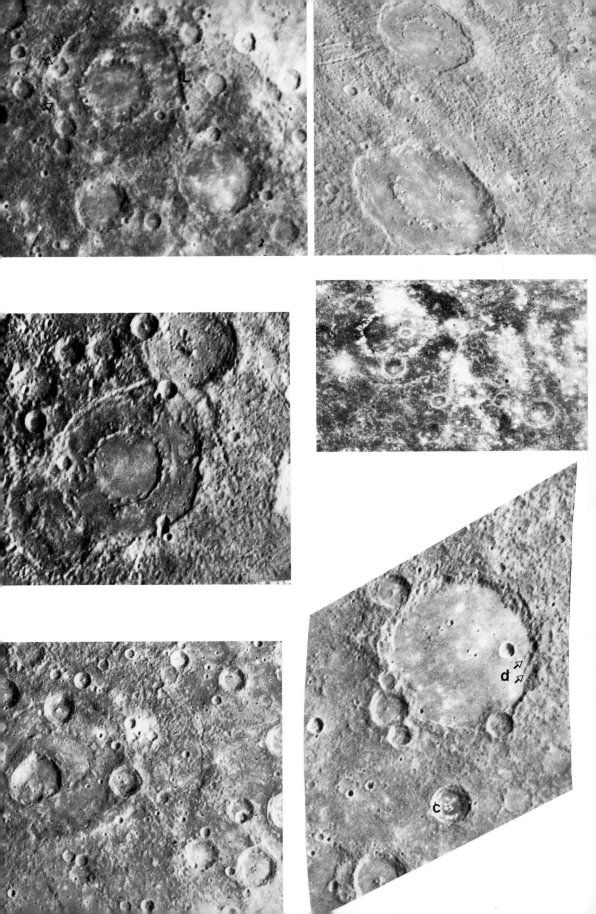

Mercury: The Intercrater Plains

The Intercrater Plains unit is the most widespread geological-geomorphological unit on Mercury. As the name implies this unit is seen between large craters; smaller craters, less than about 20km, are included in the unit simply because they were too small to represent on the original geological maps. The Intercrater Plains are thought to have been present before most of the craters because the ejecta from the craters has excavated secondary craters on the plains surface; also, Mercury's high gravity has restricted the ejecta blankets quite close to the craters, leaving much of the earlier surface uncovered. The Intercrater Plains accordingly represent a stage in the planet's history when the record of virtually all earlier meteorite impacts was erased. This was probably achieved by very extensive volcanism; that is to say, the Intercrater Plains are volcanic, although it is possible they result from wholesale melting of the crust at this time — during the differentiation into core and crust.

The resultant terrain is like that in the top picture. Most of the small craters visible here are secondaries. These are often elliptical and some are clustered together in crater chains. They are also relatively shallow, and all of these characteristics can be used to distinguish them from small primary craters which are circular and (unless rather old) sharper and deeper. The large extent of broad, level outcrops such as this (250km in this picture) shows that these plains are not merely the degraded ejecta blankets of surrounding craters and basins.

While half of the photographed hemisphere is dominated by the Smooth Plains in and around Caloris, the other face presents a fairly homogenous expanse of Intercrater Plains, and the indications are that the rest of the planet mirrors the known half (see page 68). This would explain Earth-based observations that one half of Mercury is more 'contrasty' — the half with Caloris — while the other is 'blander'. An example of the Intercrater Plains on the bland hemisphere is shown in the lower photograph. Typically the terrain is fairly level but pock-marked by many small secondary craters. In this area is a 'ghost' crater (A) which has been filled to the same level as the surrounding Intercrater Plains, but by material which is less cratered and therefore younger than the surrounds, although not enough to be Smooth Plains terrain. Furthermore, this material is exactly the same colour as the surrounding Intercrater Plains, while Smooth Plains generally show at least a slight colour difference. Crater B, adjacent to the filled crater, is quite devoid of any fill. Clearly, then, not all Intercrater Plains are of the same age. Such observations are usually taken as strong evidence of a volcanic origin for such plains.

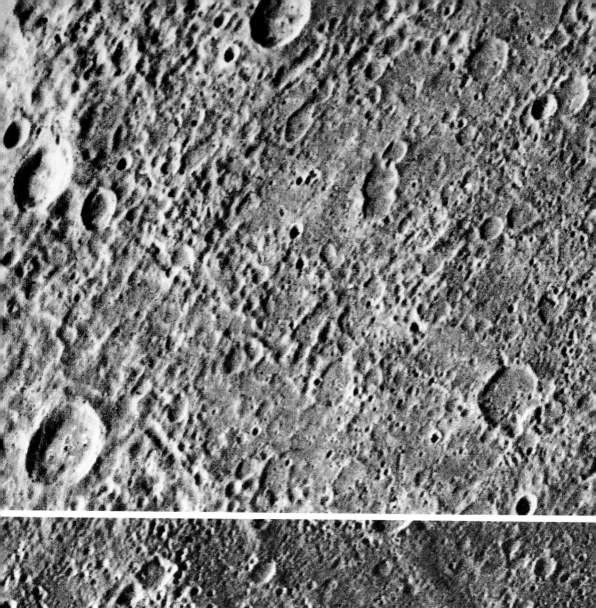
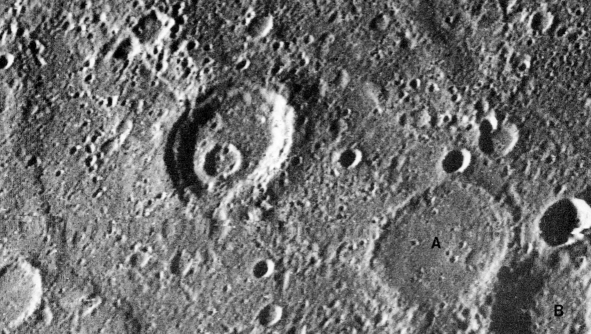

Mercury: The Earliest Mercurian Surface

Beneath the craters and basins of the highlands of the Moon are yet more craters and basins — heavy meteoritic bombardment early in the history of the Solar System has largely destroyed the original crust. Beneath the craters and basins on Mercury, however, are the Intercrater Plains. The question arises as to whether this unit represents the original mercurian crust or whether it overlies the original crust, saturated with overlapping craters like that of the Moon. If it represents the original crust then it might have been formed by the accumulation of light silicate crystals, such as feldspar, at the surface of a molten or partly molten planet. At the same time, metallic iron and nickel would have sunk to form a core. Such a differentiation process has been proposed as being responsible for the anorthositic (largely feldspar) lunar highlands and may have been important early in the history of the Earth. This process would have continued longer on Mercury than the Moon because, generally speaking, the larger the planet the longer it retains its heat. This would explain, if the Intercrater Plains constitute the original mercurian crust, why it has less craters in it than the lunar highlands — most of the meteorites impacted before the crust was formed.

However, although it is possible that the early crust was more plastic and less capable of preserving large craters, there are some indications that the Intercrater Plains are volcanic in origin. Admittedly the plains would have to be very thick, because they are very level and extensive and show little indication of an underlying heavily cratered terrain. The central picture opposite shows that the Intercrater Plains did not all form at the same time, since a very old crater (rim arrowed) can be seen to contain plains indistinguishable from those outside its rim. In fact, the whole succession — exterior plains, then large crater impact, then interior plains — has been eroded to such an extent by secondaries and small primaries that a virtually homogenous surface has been produced. The extended terrain-forming episode indicated by such superpositional relationships suggests a volcanic origin for the plains.

The rocky knob (C) in the top picture has been likened to a volcanic edifice or, alternatively, to a spur of rock from a rugged terrain underlying the Intercrater Plains; in either case it is evidence in favour of a volcanic origin for these plains. Other rocky knobs — for example, those in the bottom right picture (k), also protrude from other parts of the mercurian surface, possibly indicating that there is indeed a rugged crust underlying the Intercrater Plains. Several low dome-shaped features are also of volcanic origin.

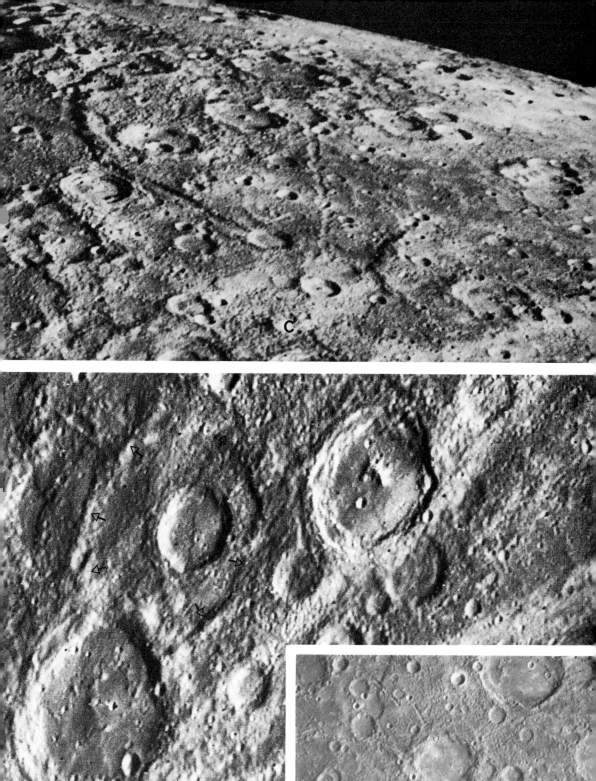

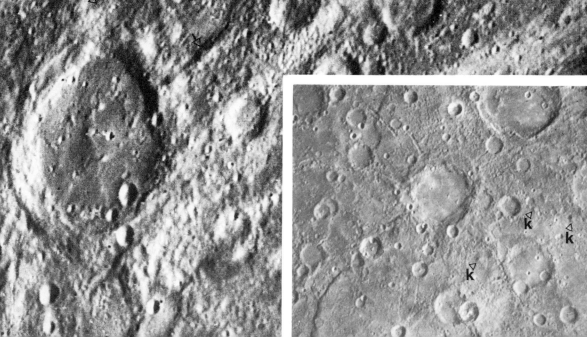

Mercury: Smooth Plains

Perhaps the greatest obstacle to the proponents of a volcanic origin for the Smooth Plains is the lack of any clear volcanic features. There are certainly no great volcanoes like those on Mars, at least on the photographed hemisphere. The problem is that the highest-resolution pictures are too few to pick up small and localized volcanic features such as the domes or lava flow-fronts found on the Moon. There are, however, larger features consistent with volcanism which can be seen. For example, among the most intriguing types of feature quite commonly found on Smooth Plains are large rimless pits like those in Caloris (bottom right). Such features are found on the Moon where they are accorded a volcanic origin (see page 44) and, although the mercurian pits are generally larger than the lunar ones, it is likely that they were caused by surface collapse or subsidence, very possibly associated with volcanic activity.

Other sources suggest that there is a scarcity of positive relief volcanic features because most of the lava flows were fed from intrusive dykes. Several examples of these possible dykes are shown (arrowed) at lower left, intersecting the rim of a large basin. The 'dykes' do not deviate when they intersect the topographic high of the rim, transectional behaviour which is strongly indicative of a vertical structure.

The stratigraphic relationship between the Intercrater Plains and Smooth Plains is shown clearly in the upper left picture where the older more rugged unit (I) dips beneath the smoother, younger terrain (S). Occasionally the rim of a buried crater can be seen protruding through the overlying Smooth Plains, which in this case lie in a region near the north pole of the planet.

Unlike the Moon, where there are dark *maria* and lighter highlands, there are no very sharp colour differences between Smooth Plains and the Intercrater Plains nor between either of these and the cratered terrain, except where the latter are fresh and rayed. Sometimes Smooth Plains appear lighter than the surrounds because small craters seem to produce brighter ejecta on their surfaces. An example of this is shown in the upper right photograph where a clear boundary (arrowed) is evident between the Smooth Plains to the north and Intercrater Plains to the south. We see also a curve of rugged terrain (S) which may be the remnant of an ancient basin rim, perhaps 800km across.

This last photograph has been processed by NASA's experts to enhance topographical detail to the maximum. It is called a Modulation Transfer Function (MTF) picture.

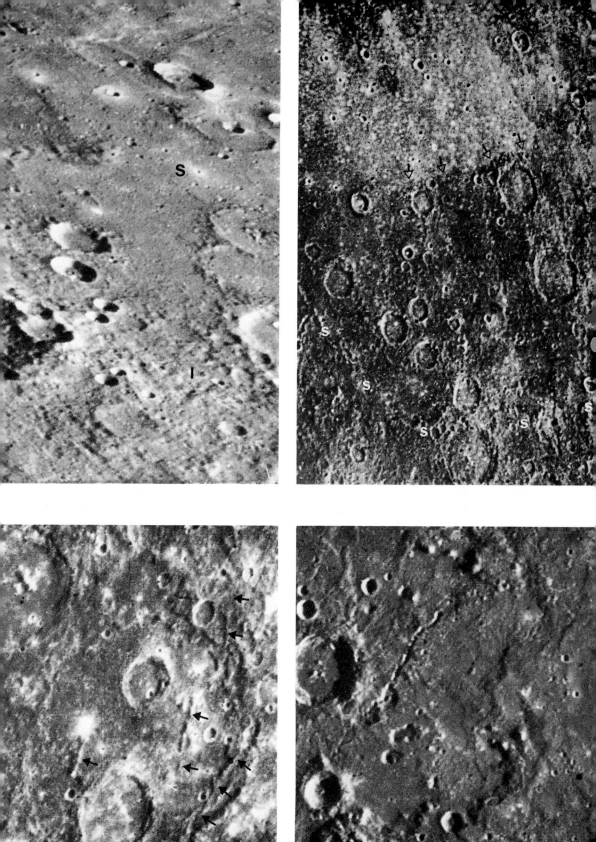

Mercury: Tectonics

Mercury, in common with the Moon and Mars, does not show any convincing evidence of plate tectonics having taken place at any stage in its history. It is thought that this is because these three planets are smaller than the Earth and so cooled more quickly, and perhaps because their crusts were too thick for plate tectonics to occur.

Instead, on Mercury, when the core or crust began to cool and contract, the compression produced countless low-angle thrust faults which show up on the surface now as long winding escarpments. Most of the scarps appear at first to be randomly distributed, although some scientists claim to detect definite trends within the scarp population. One of the largest of these scarps is shown far right. Discovery Rupes, as it is called, is hundreds of kilometres long and perhaps 3km high at its highest. There is some dispute as to how many of these features occurred before or after Caloris, but this one at least appears to be post-Caloris by virtue of the fact that it transects Smooth Plains within two craters, the larger of which is about 60km across. The scarp in some places has a terraced look suggesting it is composed of several fault planes. In the crater is a smaller scarp facing Discovery Rupes. The crater appears circular in outline, and it has been estimated that it requires several kilometres' movement on such a fault before there is any noticeable effect on the circularity of the crater. Even so, other craters are seen to be quite noticeably affected by scarps, a key factor in determining their nature.

Several scarps are shown in the top left photograph. One in the upper right hand corner of the picture, seems to separate darker (d) and lighter (l) plains within the crater whose floor it crosses.

Scarps such as that (a) to the left of the picture, which transects the Smooth Plains within a crater floor, have been suggested as lava flow fronts, but the viscosity of lava required would be extremely high. In any case, they do not look much like lobate flow fronts and, furthermore, can often be followed beyond the crater, showing them to be tectonic. This particular scarp can be seen cutting one wall of the crater, continuing across the middle in the Smooth Plains but oddly detouring around the central peak, then up the opposite wall into the Intercrater Plains where its continuation (b) can be seen to transect the terrain (shown at higher resolution in central picture) in such a way as to indicate a near-vertical fault.

The picture at lower left shows a relatively uncommon ridge which, like most scarps, was also caused by the crumpling of the crust.

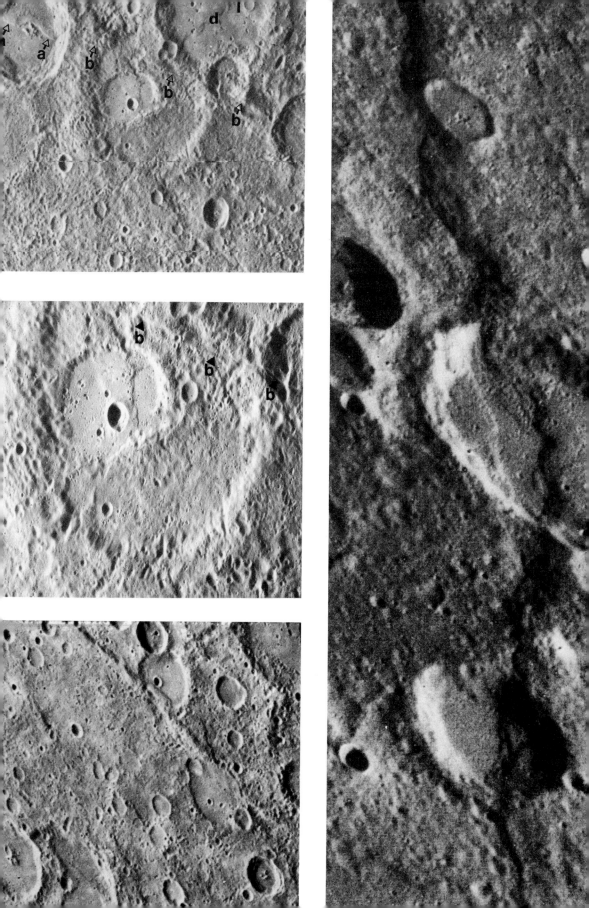

Mercury: Lineated Terrains

As well as curved and sinuous fault escarpments there are many straight scarps suggestive of near-vertical normal faults. In some cases these concentrate in generally old regions called Lineated Terrain. An example of such a unit is shown in the upper picture opposite. The lineations in this area trend approximately southwest to northeast, cutting very old Intercrater Plains and cratered terrain, and themselves appear rather degraded. In some cases it looks as if two lineations might bound a down-faulted graben (G); if this is the case the graben must surely have formed long before the compressional thrust fault scarps. The lineations in this picture are in some places radial to the basin Tolstoj, a corner of which (dotted line) can be seen in the bottom left of the picture.

In an area around the Caloris antipode — the part of the planet exactly opposite Caloris — linear grooves cut through a hilly region unlike anything else seen so far on this or any other planet (lower left photograph). The lineations in this region, termed the Hilly and Lineated Terrain (although known informally as the Weird Terrain), trend east-northeast and west-northwest (north is to the right of both the lower pictures). This is exactly the same direction as the linear crater wall segments illustrated on page 79, and they are thought to have the same origin. That is, they are thought to have originated as a stress pattern in the crust, possibly manifested as strike-slip faults, formed in response to the slowing down of the planet's rotation under the effect of tidal interactions with the Sun. This stress pattern probably exists in much of the mercurian crust, where it has influenced the formation of everything from craters to later faulting — and probably including intrusive dykes. It has been suggested that the east-northeast/west-northwest pattern is particularly clear in this region because the faults were reactivated by huge seismic waves focussed through the planet from the Caloris impact. The effects of Caloris must have been cataclysmic, comparable no doubt to those wrought upon the Moon by the Imbrium planetesimal which, it is thought, would probably have broken the Moon into pieces had it been very much larger. Similarly, the martian moon Phobos could not have sustained a much larger impact than that which caused Stickney.

At lower right is a higher-resolution picture of part of this area. It is clear that the lineations have affected the small crater C (diameter 35km) in the middle of the picture, but have not affected the Smooth Plains within it. Therefore, the lineations were formed after the small crater but before Caloris (assuming all Smooth Plains are post-Caloris).

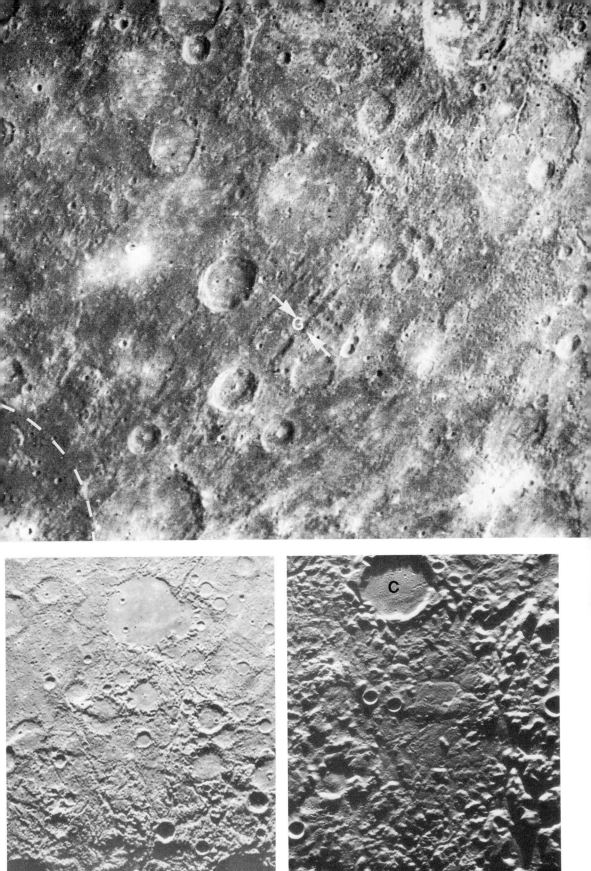

Mercury: The South Polar Region

It is a relatively easy matter to present interpretations and speculations on features and terrain which have been extensively discussed by scientists involved in research in this subject, and using the best examples — as we have done in these pages. It is a different matter to apply what is known of mercurian geological history to a particular area where everything is mixed in together; as, for example, in the region shown opposite. This picture is a photomosaic composed of several photographs of about the same resolution. It covers an area approximately 2,500km long and 1,400km wide, near the south pole (SP) of Mercury.

The main unit, and the oldest, comprises the Intercrater Plains. Superimposed on this are many craters and basins which have made the surface of the Intercrater Plains very rough, except in the central part of the area: here there seem to be far fewer large impacts, and this may explain why the surface is much less rugged; or it may be that the plains of this region are slightly younger and have covered any large craters or basins. As is typical on Mercury, the rugged Intercrater Plains and the smoother unit merge into one another imperceptibly, making both geological mapping and interpretation difficult.

Sometime after most of the large basins and all of the Intercrater Plains had been formed, the crust began to crumple, forming the ridges and scarps which concentrate in the smoother part of the region; this crumpling may have been caused by the planet's core or crust cooling down and contracting.

About this time, Caloris was formed — an important event, since it was of very short duration, enabling geological history of Mercury to be divided into two time groups, pre-Caloris and post-Caloris, even in areas on the other side of the planet from the impact, such as that shown opposite.

Smooth Plains production then commenced and — as far as can be determined — within a relatively short while ceased. Since that time, probably no less than three billion years ago (by analogy with the Moon), the only substantial events on the surface have been the, generally small, post-Smooth Plains impacts (such as i, lower right), the most recent of which have left extensive ray systems, as at the top of the picture.

Mission scientists have shown that the Smooth Plains tend to be slightly darker and redder than the surrounding terrain, and that fresh craters and their rays tend to be slightly bluer. An interpretation which they consider consistent with these colour and albedo relationships is that the Smooth Plains resemble the low-titanium low-iron lunar *maria;* and that the mercurian crust itself is in general low in titanium, metallic iron and iron oxides — that is, more like the lunar highlands than the lunar *maria.*

Mars: An Astronomical View

Mars, often called the Red Planet because of its colour, has about half the diameter of Earth and lies in an orbit outside that of the Earth. Although the telescope was invented in about 1608, it was not until 1659 that telescopes had improved enough for recognizable drawings of Mars to be made showing it to have light and dark areas. With careful observations, astronomers gradually built up a body of knowledge about Mars. The dark features were charted and named in the nineteenth century, and it became clear that these dark areas changed in their degree of darkness and often in their size. It was also noticed that Mars had icecaps at the poles. Mars, like the Earth, has an inclined axis of rotation resulting in seasonal changes in the length of night and day as the planet revolves around the Sun, and thus the icecaps advance and retreat with the seasons.

As early as 1784 the British astronomer William Herschel was of the opinion that Mars had a 'considerable atmosphere'; but it was not until 1947 that Gerard Kuiper, in Texas, showed that the atmosphere of Mars consisted essentially of carbon dioxide. Most of the observations of Mars' atmosphere made from Earth inferred too high an atmospheric pressure, and it was not until Mariner spacecraft flew by Mars that the atmosphere was determined to be a thin one — it has only a hundredth the atmospheric pressure of Earth.

Astronomical observations showed also that the martian atmosphere sustained clouds. Several types of clouds were observed: these were either white clouds of minute water crystals or droplets, or yellow clouds of fine dust particles swept up from the surface by high winds.

Whenever Mars is mentioned, most people's minds turn to thoughts of canals. Canals on Mars were first documented by an Italian astronomer Schiaparelli who, like others before him, thought he saw thin dark lines crossing the surface of the planet. The idea that these lines might be canals was taken up by the astronomer Percival Lowell, who built a telescope at Flagstaff in Arizona to study these features further. He became convinced that the canals were evidence for intelligent beings and constructed a fabulous picture of a martian civilization irrigating the land by canals leading from the icecaps. Many astronomers doubted the existence of canals and explained them as artifacts of observation, a view supported by spacecraft observations which have demonstrated that the canals do *not* exist. Nevertheless, Lowell's ideas have made a paradoxically important contribution to our science in helping to foster the idea of life on Mars—an idea that culminated in the Viking mission, which was expressly designed to look for life; so far, it appears, without success.

The picture opposite shows a view of Mars taken by the Mariner 7 spacecraft in 1969. This view is substantially better than can be obtained by Earth-based telescopes and shows the dark markings and a prominent polar cap.

Mars: Early Missions

With popular opinion being coloured by Lowell's views of a pleasant and almost Earth-like atmosphere on Mars, and the opinion of many scientists that Mars was essentially a large desert as evidenced by the many observed duststorms, it came as something of a surprise when the first missions to Mars—those of Mariner 4 in 1965 and Mariner 6 and 7 in 1969—showed the surface to be covered by impact craters. If Mars had been an active planet with erosion more akin to that on Earth than that on the Moon, the ancient impact-scarred surface should have been removed long ago. It should be said that a few people had predicted impact craters, notably that great supporter of the impact theory, Ernst Öpik of the Armagh Observatory in Northern Ireland.

An example of a Mariner 7 picture of the cratered surface of Mars is given opposite above. The largest crater in the picture is about 240km across. Pictures such as this were interpreted by scientists to suggest that the martian surface, like that of most of the Moon, represents a fossil surface formed early in the history of the Solar System when the planets were being bombarded by large meteoritic and asteroid-sized bodies. In general the craters are more degraded than those on the Moon, and this was thought to result from slow erosion by wind action.

However, scientists working on these missions noticed that in certain areas there was a new kind of terrain which they called 'chaotic'. An example of this is given opposite (top right) and shows vast areas of collapsed and jumbled terrain. Chaotic terrain was thought to represent the beginnings of internal activity within the planet, and some scientists even went so far as to suggest that Mars had only just reached the point where it was hot enough to generate internal activity.

The first missions to Mars looked at the southern hemisphere, and it was only when Mariner 9 went into orbit around Mars in 1971—arriving during a duststorm so that the craft was unable to see the surface for several weeks—that it was realized that, although the southern hemisphere was densely cratered, the northern hemisphere had relatively few craters. The lower two pictures show one of the first features to be observed from Mariner 9 when the duststorm cleared. (These pictures illustrate the quality of Mariner 9 photographs.) Here we see the great volcanic mountain of Olympus Mons rising more than 20km above the surrounding terrain. The one on the left shows the whole volcano, which is about 600km across, while the one on the right shows details of lava fields.

This volcano was just the first of many relatively young features to be observed on Mars, including large fault systems, comparatively smooth plains and extensive fields of wind-formed deposits that make up the dark patches observed telescopically. Mariner 9 saw also winding channels, apparently evidence of flowing water at an earlier period in Mars' history when the atmosphere may have been different from that of today.

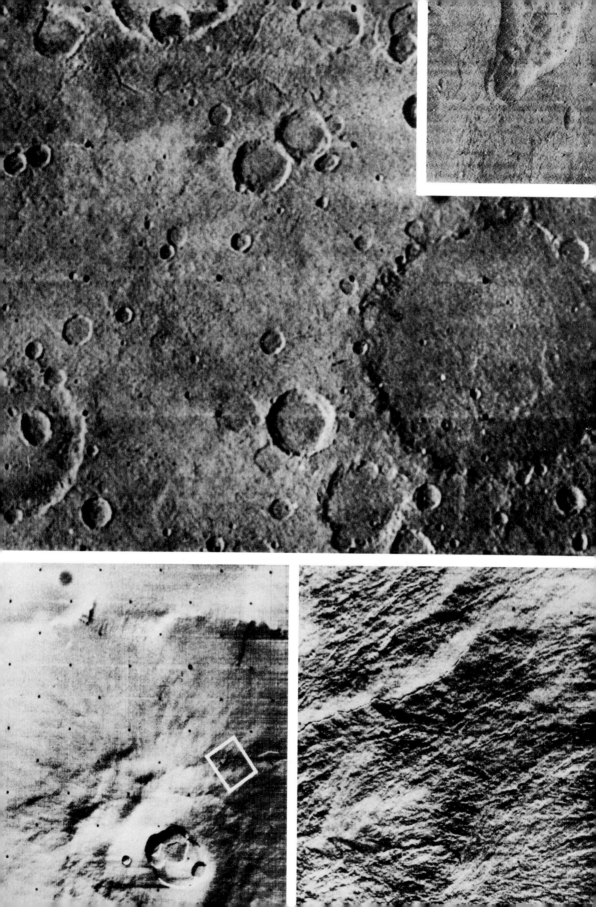

Mars: Mariner 9 View

Mariner 9 was intended to be one of a pair of spacecraft that would go into orbit around Mars; unfortunately Mariner 8 failed on launch and so Mariner 9 was left to do the work of two. It is to the credit of the engineer and scientists that this mission operated for a whole year, and using a highly inclined orbit around Mars was able to image virtually all of the surface of Mars. By the end of the mission there was enough data to produce topographic maps covering the whole of Mars and to establish a framework for the geology of Mars. From the equatorial topographic map (shown opposite) compiled by the US Geological Survey, the main geological features of Mars can be observed.

The southern heavily cratered hemisphere is clearly seen. The boundary between it and the relatively poorly cratered northern hemisphere forms a great circle that is slightly oblique to the equator. The density of cratering in the south varies from place to place, suggesting that in some parts even the southern hemisphere has been modified by later activity, some of the craters becoming buried. Two large impact basins are particularly prominent, those of Argyre, some 1000km across, and Hellas which is even bigger. Modifying agents of the southern-hemisphere cratered terrain include wind action, burial by later deposits, probable water erosion, and volcanic activity which has covered large areas, particularly in the region of the Hellas basin.

The dichotomy between the northern and southern hemisphere is paralleled by the overall global form, there being a bulge in the southern hemisphere where the terrain is several kilometres higher than the plains to the north.

It was the relatively smooth terrain of the northern hemisphere that attracted those who were planning the Viking mission in 1976. The aim of this mission was to put two orbiters around the planet and two landers on the surface; each of the landers was equipped with sophisticated instrumentation to look primarily for evidence of life on Mars, and also to analyse the surface soils, photograph the terrain around the landing site and make meteorological observations. When Viking eventually arrived at Mars in the middle of 1976, photographs taken by the orbiters demonstrated that, although the northern hemisphere had looked smooth from Mariner 9, the surface was rougher than had been expected. The 'smooth' plains of the northern hemisphere had many small impact craters on the surface, were cut by small channels and had numerous irregularities such as small hills that were potential hazards to the landers. After much hard work by the Viking project members two landing sites were chosen. One of them was in the Chryse basin, an embayment into the southern cratered plains just south of Acidalia Planitia; the other was further north in Utopia Planitia. Once down, these two landers were to give us an enormous amount of information about the surface conditions on Mars while the orbiters provided pictorial coverage of the surface and the clouds. Other experiments included infrared spectroscopy. It is from the high-quality Viking data that we illustrate most of the section of this book dealing with Mars.

Mars: A New Style of Impact Crater

Very few of the martian craters seen in Mariner 9 pictures are fresh: the large craters, more than a few tens of kilometres in diameter, are nearly all eroded so that features on the ejecta blanket are destroyed; and the resolution of Mariner 9 was not sufficient to identify details of the form of smaller craters. The first pictures sent back by the first Viking Orbiter clearly showed that not only were there many small craters but that the majority of these were relatively fresh and thus presented a hazard to the Viking landers because of the possibility of large numbers of ejecta blocks lying on the surface around them — if the craters had been less fresh, boulders and blocks would have been largely obliterated by erosional and depositional processes.

As we have seen in our discussions of Mercury, gravity strongly affects the distribution of the ejecta around a crater. For Mercury, with a surface gravity 2.3 times greater than that of the Moon, the ejecta lies much closer to the crater than it does on the Moon. It so happens that the surface gravity on Mars is almost exactly the same as that on Mercury, and many workers felt that the original form of martian craters would be the same as those observed on Mercury. However, pictures such as the ones opposite, photographed early in the Viking mission, show that this is not the case. Although the interior of these impact craters is similar to those seen on both Mercury and the Moon, the ejecta spreads outside the crater are quite different. The outer part of the continuous ejecta is marked by a low, almost continuous ridge, and it is the presence of this ridge that has led to the name 'rampart crater'.

Inside this low ridge the ejecta is apparently thin and has a hummocky, ridged and striated surface. Outside the rampart there are hummocky and dune-like features of discontinuous ejecta and, in some regions, a few scattered secondary impact craters can be seen.

The photograph at bottom right shows a crater where not only does the ejecta have a rampart front but also the edge of the ejecta is lobate, having the form of a series of overlapping flow units. It is craters like this that provide good evidence that the ejecta, once emplaced close to the crater, flowed outwards to well beyond the range expected for ejecta thrown out ballistically under the surface gravity of Mars. It is worth noting that these sheets of ejecta appear to have been obstructed by the raised rim of a nearby crater. Also, the primary impact crater was formed close to a pre-existing smaller crater. The older crater has therefore acted as a barrier and flows are diverted around it.

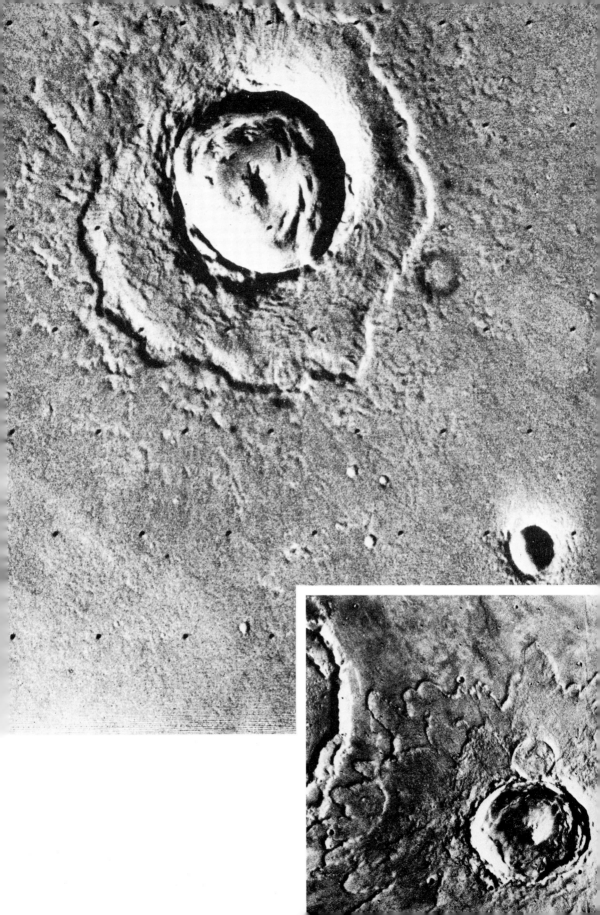

Mars: Arandas

The two Viking pictures opposite show Arandas, a 28km-diameter crater lying on the northern plains near the border with the cratered southern hemisphere in Acidalium Planitia. The interior is like that of many craters of this size on the Moon and Mercury, showing an upper scarp of bedrock around the rim and below this a slope of talus. In the middle is a well defined central hill formed by bedrock being thrust up into the floor of the crater during the final stages of impact.

Immediately outside is a steep rim formed by the bedrock being pushed up and outwards on the edge of the crater of excavation. Around Arandas, as with many other martian craters, there are two tiers of continuous ejecta: an inner tier, illustrated by the main picture opposite; and an outer tier, the edge of which is shown at top right. It appears that two distinct layers of ejecta flowed or slid outwards from the crater. On the inner tier of ejecta we see well defined radial striations typical of certain types of landslides on Earth and resulting from differential movement within the slide material, producing shear zones parallel to the direction of movement. The edge of the inner tier is lobate, and there are ridges concentric with the edge that may be considered as flow-ridging in a viscous material. So all the evidence suggests strongly radial sliding of the ejecta along the surface after it was deposited.

On the left-hand side of the main picture, near the edge of the inner tier, is a small crater that was there before Arandas was formed. Looking closely we see ridges formed on the 'upstream' surface of the ejecta, and it seems the small crater has acted as an obstacle to the forward movement of material. Because the ejecta has not filled the crater we may infer that when it was emplaced its thickness was less than the height of the crater's rim; thus it was emplaced as a dense sheet rather than as a finely divided cloud of material that later came to rest. This is consistent with our idea of a rockslide type of mechanism. The outer tier has similar features, although the edge of the ejecta is distinctly lobate. This again suggests an outward movement of the ejecta.

Why are these craters so different from those of the Moon or Mercury? One difference is the presence on Mars of volatiles, in the form of an atmosphere and also quite probably as permafrost. Volatiles could assist in the forward sliding of ejecta. When ejecta hits the surface it still has a forward component of movement. There are several ways in which volatiles could have caused the ejecta to continue moving forward: one possibility is that, as the ejecta hit the surface at high velocity, it vaporized the permafrost and at the same time entrapped the gas below the rapidly falling rain of material; the gas attempted to escape upwards through the ejecta, fluidizing it and assisting it to move forward.

Mars: Other Impact Crater Forms

Although most of the small craters on Mars are of the rampart type, some are strikingly different. Opposite we see in the main picture a crater of 18km diameter situated on the edge of the Kasei Vallis on the western side of the Chryse basin. Viking has shown that there are a small number of craters similar to this elsewhere on Mars. The striking feature is that they do not have the rampart structure but instead have a well defined continuous ejecta dominated by radial striations. Immediately outside the crater lip are some lobate features that are reminiscent of those on the rampart craters, but nevertheless the total appearance of the ejecta blanket is quite different. The inner lobate forms on the ejecta appear to be about 20m thick, and it will be noticed that they tend not to have such a strong radial pattern imposed on them. They form an imbricate pattern and were almost certainly formed by avalanching or flow of ejecta down the rim of the crater in much the same way as we see flows on the rims of lunar craters such as Tycho. The main extent of continuous ejecta is radially marked, and the origin of these radial striations is not entirely clear. There is the possibility that they have a similar origin to the radial features seen on the rampart craters, but alternatively they may have been formed by erosion caused by scouring of the surface by radial flow of material from the main crater. The discontinuous ejecta also is radially striated.

Craters such as this one are enigmatic and suggest that not all the craters on Mars formed under the same conditions. Possibly some were formed in areas where there was no permafrost at a time when conditions elsewhere on Mars were different; however, there are, in the same region, typical rampart craters.

It is interesting to note that this crater must have formed after the Kasei Vallis was carved out, because the ejecta pattern is seen to cover the scarp on the edge of the valley shown in the top left of the picture.

Another type of small crater seen on Mars is that illustrated at the top. These craters are similar to those seen on Mercury, with the ejecta piled up close to the primary craters. Mercurian-type craters occur in specific areas, notably relatively fresh lava flow surfaces. Lava flows tend to be fairly dense, pore-spaces being restricted to joint cracks and open vesicles, and it is likely that within some of these flows there was no permafrost. Under these conditions, given that the surface gravity on Mars is similar to that on Mercury, we might expect small martian craters to be similar to those on Mercury and thus we might say that the form of impact craters gives us an indication of the nature of the bedrock underneath the craters.

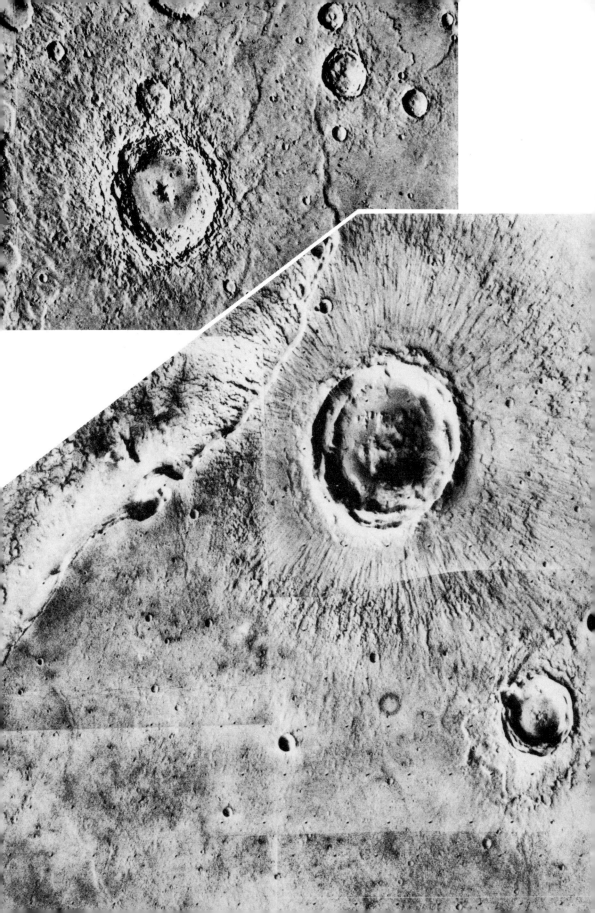

Mars: Degradation of Martian Craters

It was discovered from studies of the Moon that a useful way of determining relative ages of planetary surfaces is to compare the degree of degradation of the oldest craters on one surface with those of another. Surfaces that have more degraded craters on them are considered to be older, having been exposed to the erosive processes longer. This is not a particularly reliable method on Mars. On the Moon or Mercury the main erosive process is impact, which is a relatively uniform process over the whole planet at any specific time; on Mars we have the possibility of a number of different processes of erosion taking place and these may occur at different rates on different parts of the planet. For example, wind erosion may be more prominent in some regions than in others, and clearly in the vicinity of channels water erosion has more strongly affected the local terrain than elsewhere. Nevertheless, it is important to understand the chronology of erosion of an impact crater on Mars. To do this it is necessary to know the original form of the crater and its surrounding ejecta. Because martian craters are so different in initial form from those seen on Mercury and the Moon we cannot assume that they follow the same degradation pattern. The picture opposite (top left) illustrates a small crater that has been eroded to some extent. Craters of this type were known from Mariner 9 pictures, and it was assumed that they had in their original form looked like lunar or mercurian craters. The wide belt of ejecta around the crater was considered to result not from ejecta emplacement but from differential erosion: blocks lying on the continuous and discontinuous ejecta acted as an armour-plating to the surface so that less erosion occurred in the vicinity of the crater than elsewhere. The material around the crater was therefore considered to be ejecta and bedrock left behind after the erosion process, and could not be taken as indicative of the original extent of the ejecta. This view resulted entirely from the fact that martian craters were thought to be similar to those on other planetary bodies.

With the higher-resolution Viking pictures it is clear that, although the crater illustrated is eroded, its present form is not far removed from the original form: we can see the inner and outer tiers of ejecta and only a small amount of erosion has taken place, mainly on the edges of the outer tier. Such craters are therefore relatively young.

The larger craters (main picture) on Mars tend to be older and very little of the continuous ejecta remains, there usually only being left a battered rim of a crater which (in the fresher ones) still preserves the inner terracing. The large crater on the left in the main picture has clearly been eroded and we can only just recognize the original texture of the ejecta blanket; the inner terraces are still partly preserved. With craters older than this no ejecta blanket is seen and the rims become progressively more battered with more superimposed craters than we see on this one.

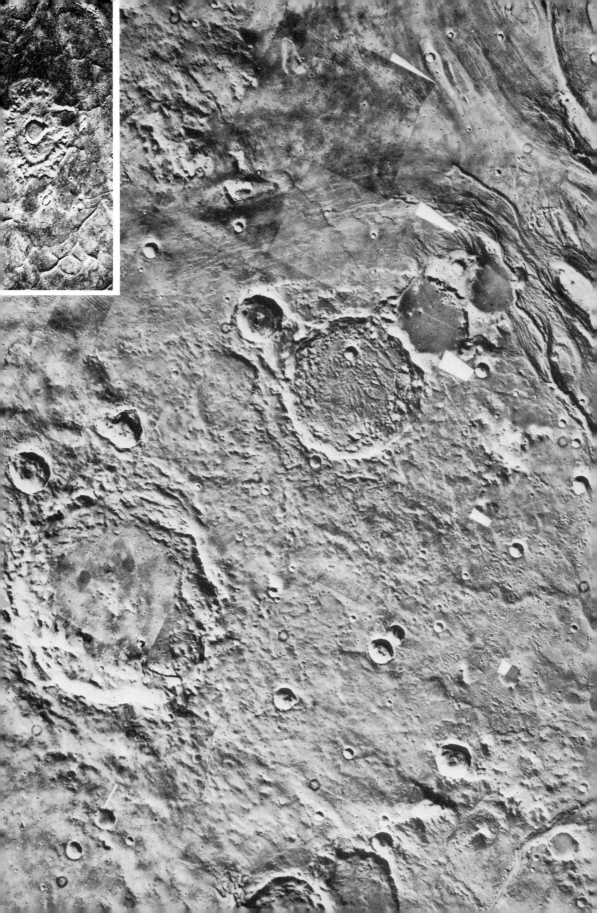

Mars: Hellas Basin

Like the Moon and Mercury the surface of Mars includes a number of enormous multi-ringed depressions or 'basins' of probable impact origin. More than twenty have been recognized so far. The largest martian basin, indeed the largest known in the Solar System, is Hellas, illustrated opposite. This depression, 1,600km across and 5km deep, is surrounded by a much eroded and partially buried mountain rim, and traces of four more concentrically arranged 'rings'. It has a complex interior floor called Hellas Planitia lying some 2-4km below the martian datum. The rim of a much smaller basin is arrowed. From the large number of superimposed craters, this small basin is considerably older than the interior of Hellas.

Hellas is clearly seen through the telescope from Earth and is often one of the brighter regions on the planet's disk, sometimes having been mistaken for the south polar cap. This observed brightness may be caused by clouds or even frost. It has also been observed telescopically that major duststorms start near Hellas.

It appeared from pictures taken early in the Mariner 9 mission to Mars that Hellas had an almost featureless interior. This was merely the result of persistent duststorms concealing the ground beneath. A considerable complexity of terrains was revealed later and during the Viking mission when there were fewer duststorms in the region.

Knobs, scarps and furrows are all prominent. Some of the interior, at R for instance, is occupied by volcanic plains crossed by ridges like those of the lunar *maria*. Much of the basin floor has an etched and scoured appearance, testimony, like the characteristic duststorms, to considerable past and present wind activity. Indeed Hellas behaves like an enormous dustbowl.

Basins are important, not merely because of their visual prominence. Their structural influence on planetary crusts is pervasive. Two large volcanoes, Amphitrites Patera and Hadriaca Patera, seem to have developed on fractures concentric to Hellas. (Part of Amphitrites Patera is visible at A.) Their lavas have breached and buried the southern and northeastern rims of the basin.

As so frequently on Mars, the oldest preserved features in the area of Hellas are small and moderately sized impact craters. The Hellas basin postdates the late stage of planetary bombardment these craters represent. Subsequent cratering, and probably considerable water and wind activity, have eroded much of the initial basin relief. The volcanic episodes which produced Amphitrites Patera and Hadriaca Patera came much later. They have been succeeded only by local fluvial activity (note the channels at C, for instance) and the continuing effects of wind activity.

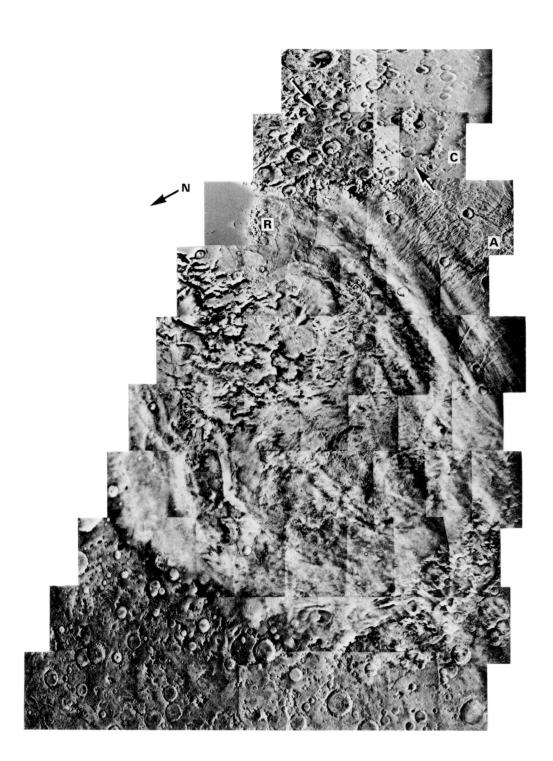

Mars: Argyre Basin

Argyre basin is much younger in appearance than Hellas, with a more rugged and continuous rim and fewer superimposed craters. The picture opposite gives an oblique view of the basin's eastern rim and interior. It also includes the small basin Galle (G). Above the horizon several haze layers in the atmosphere are clearly visible.

Argyre consists of an interior plain some 900km across surrounded by a broad ring of mountains. Several other rings are discernible, although the outermost are preserved only as low scarps to the northwest of the basin centre.

The outer regions of the basin include numerous radial valleys and arcuate troughs similar to those of the lunar basin Imbrium (which is of similar size). On the other hand, no ejecta deposits or secondary craters have been detected around Argyre, unlike the case with Imbrium.

It is striking how the dark plains materials, lying between many of the craters of the southern hemisphere of Mars, terminate abruptly around the basin. If this plains unit is old, the arrangement is readily accounted for: the basin-forming impact would have ruptured it. If the intercrater plains are younger than the basin, it is necessary to invoke continuing vertical motions around the basin, or enhanced erosion in the region of the basin rim, to explain their absence from the basin interior. In places intercrater plains appear to overlie the mountains forming the rim of Argyre, and certainly there are few signs that they themselves have been buried under Argyre ejecta. Both these lines of evidence suggest that these plains do post-date the basin.

Looking closely at the basin rim, the trends of the basin rings can just be made out. Inside the inner ring can be seen some of the plains materials of the basin interior. The irregular depressions at O may be wind deflation pits, which might indicate that the materials involved are rather poorly consolidated. The sinuous ridges at R may be lava tubes, dykes, or compressional features in flows. The smooth plains at S, lying between the rings, may, by analogy with lunar basins, be explained as volcanic flows emplaced along basin fractures.

Mars: Isidis Basin

Isidis is much the most modified of the three big basins of Mars. The rim, 1,400km in diameter, has been partially buried by plains materials. Isidis has also been the least well imaged by Viking of the large basins. The main picture opposite, which shows its regional setting and gross form, is taken from a shaded relief map made from Mariner 9 images.

To the south the rim is well preserved, and there are even signs of a hummocky unit which may represent degraded basin ejecta. To the north and east, the rim has evidently dropped down along faults, and been deeply buried by young plains of the Vastitas Borealis. The rim is again buried to the west, this time under the moderately cratered plains of Syrtis Major.

To the northwest, vestiges of the rim remain as swarms of small hills. This area is dominated, however, by the Nili Fossae, arcuate graben concentric to the basin. The bottom insert illustrates part of this area at moderate resolution. Graben like this have occurred where the regional southwest-to-northeast fractures of the area have been emphasized by the concentric system of Isidis. Their floors frequently appear to have been flooded by volcanic material, perhaps emplaced from the graben fractures themselves. Other volcanic features are present within this region, including irregular cones and a number of small flows (the feature at F may be such a flow).

The interior plains of Isidis have been particularly poorly imaged by Viking, with only one sequence of moderate-resolution pictures. The upper picture is taken from that sequence. The unusual ridged plains at R occupy a roughly circular region in the centre of Isidis Planitia. They appear to represent a comparatively recent episode of volcanism with material being emplaced through basin floor fractures. Some of the ridges may be lava tubes, while many are probably compressional features.

It has been suggested, by Don Wilhelms of the US Geological Survey among others, that Syrtis Major Planitia, the plains area immediately to the west of Isidis, marks another ancient basin of similar dimensions. No ejecta or rim remnants have been identified, however. An alternative explanation of the Syrtis Major depression invokes global tectonics. Syrtis Major is on exactly the opposite side of the planet from the great crustal bulge of the Syria Rise. The collapse of Syrtis Major and subsequent volcanic flooding may have been caused by the withdrawal of material during the Syria upheaval.

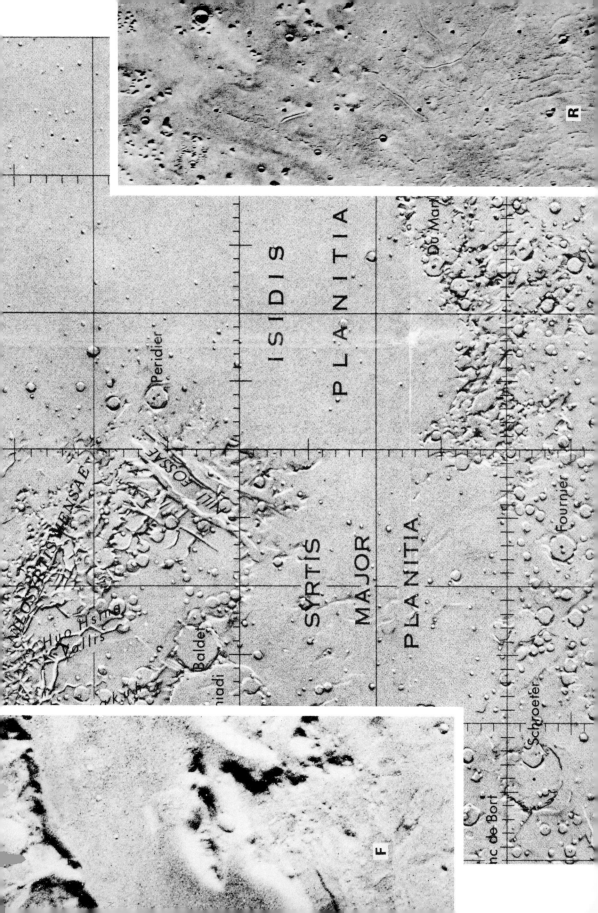

ISIDIS PLANITIA

Peridier

NILI FOSSAE

SYRTIS MAJOR PLANITIA

PLANITIA

Fournier

Schroeter

nc de Bort

Du Mar

Baldet

niadi

Hua Hsine
Vallis

NILOSYRTIS MENSAE

R

F

Mars: Other Impact Basins

The number of basins per unit area on Mars is only one tenth that on the Moon. This can be readily accounted for by the greater post-basin-formation activity of the martian surface. For instance, only one basin (Lyot) is visible within the northern plains, which comprise half the planet's surface area. Many more northern basins have doubtless been buried.

If one plots the number of basins within different diameter ranges against diameter, one finds for both Mars and the Moon, that the curve obtained does not approach that for small craters on the heavily cratered regions on either planet. The basins of both may have been produced by the same population of impacting bodies, one distinct from that which has produced most craters. This suggestion is supported by the apparently narrow time interval during which basin formation occurred.

Ringed basins on Mars occur with slightly smaller diameters than those on the Moon (around 120km on Mars, against 150km on the Moon). Smaller basins may result on Mars as on Mercury because of the higher gravity of that planet. The greater comparative frequency on Mars of larger basins is readily explained, again by the higher martian erosion and burial rates, which will have preferentially destroyed smaller features.

The most conspicuous feature of basins, their rings, characteristically occur in the same positions with respect to the main basin rim, on both the Moon and Mars. On Mars two distinct types of ring are found. The first (displayed by Kepler, Lowell, and Lyot, for example) are like lunar basin rings — decidedly mountainous with similar inward and outward facing slopes. The second type of ring seems confined to the older basins and has probably evolved from the lunar-like form through subsidence along circular fractures.

The top picture includes part of the fresh basin Lowell. Lowell has two rugged rings and a recognizable ejecta blanket. The ejecta blanket includes a number of details common to lunar basins; for example a radial valley, part of which is visible at R. Much erosion has nonetheless occurred, and what may be incised fluvial channels are visible elsewhere on the rim. The basin Kepler, shown in the middle picture, is much more poorly preserved. Considerable rim slumping has occurred (note the slump block at B), and the interior has been inundated with several plains units. The basin in the bottom picture is yet more subdued, and the rings are transitional between mountains and step-scarps.

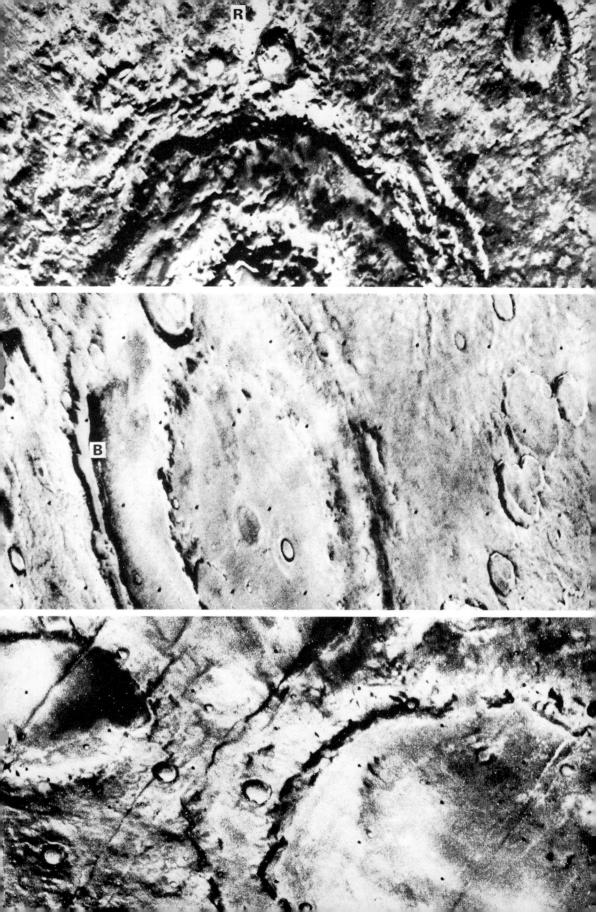

Mars: Olympus Mons

To anyone interested in volcanoes, those of Mars present a particular fascination; whereas on the Moon most of the volcanism has resulted in extensive, relatively flat-topped flows, volcanism on Mars, as well as producing extensive floods of lava, has also built large volcanic mountains. The best known of these volcanoes is Olympus Mons, the largest volcano so far discovered in the Solar System. The big mosaic opposite shows this volcano and the surrounding area. The outer margin of the mountain is marked by a scarp which is in many places as much as 4km high. Many lava flows are draped over the scarp and extend onto the plains surrounding the volcano. This is particularly clear in the bottom part of the picture. The diameter of the volcano at the scarp is about 600km, but the total diameter of the volcano including those flows that extend beyond the scarp is considerably greater than this. Estimates of the volcano's height range from 20 to 27km. Whatever the correct altitude of the summit above the surrounding plains, it is clearly more than 20km, roughly twice the height of the Hawaiian volcano Mauna Loa above the seafloor on which it stands.

At the summit of Olympus Mons is a caldera complex nearly 80km across. These calderas were formed by collapse into the central conduit up which the lavas rose. Collapse results from sudden withdrawal of lava down the conduit, removing support to the summit part of the mountain. In this picture we can see details of the flows showing that they are like basaltic flows on Earth, which normally flow with a relatively low viscosity giving them the possibility of spreading easily. This is seen particularly well in the picture at top right which shows lavas that have flowed over the edge of the bounding scarp of Olympus Mons. The fine texture represents small channels and collapsed lava tubes. It is not clear where the flows originated, as only rarely do we see lines of small eruptive vents on the flanks of the volcano, and many flows may well have originated from the summit itself rather than from vents opened up by radial fissuring on the flanks. We cannot be completely confident that these lavas are basaltic in composition, but certainly, at the time of eruption, their rheological properties were the same as those of basalts.

The age of the volcano is unknown; certainly there is no evidence of activity taking place while the volcano was being observed by spacecraft. Some of the flows in the floor of the calderas may be relatively young as they have no impact craters on them; by young we mean they may have been formed at any time during the last few million years, but, because we don't know the cratering rate on Mars, they may well be several hundred million years old.

Mars: The Olympus Mons Scarp

Several suggestions have been made for the origin of the perimeter scarp around Olympus Mons. One is that it represents a ring fault, that the whole volcano has been pushed up bodily along this fault to give a scarp all the way around the edge. Another is that, in the early history of the volcano, ash flows were erupted. Ash flows occur on Earth: large volumes of lava fragments are erupted as a dense cloud which is able to flow almost like water, eventually depositing a thick layer of ash. In the lower part of the ash layer the particles are so hot on emplacement that they weld together to give a solid rock, but elsewhere the ash is less coherent and is eroded relatively easily.

You will have noticed in the picture on the previous page that, outside the Olympus Mons scarp, there are areas of terrain which are cut by a pattern of furrows. These were thought to be the eroded remnants of the ash flows, the removal of which by erosion left the present scarp. Viking pictures throw a different light on this furrowed ground surrounding Olympus Mons.

The big mosaic opposite shows a detail of furrowed terrain; we can see that the patterns of furrows are concentric with, and radial to, the volcano and that the furrows become smaller in size towards the outer edge of the unit. One clue to the origin of this material is given by the insert, which shows the furrowed material superimposed on a pre-existing crater, half covering it. It is thus unlikely that the furrowed material is an eroded remnant of an ancient part of Olympus Mons; much more likely that it is a relatively young unit that has spread out from the volcano to cover the surrounding terrain. Perhaps the idea that these are gigantic ash flows is not unreasonable although, if they are, they are relatively young. However, another possibility is to link the furrowed material to the scarp-forming process. The edges of the furrowed material are lobate in form. This is well shown in the picture on the previous page, where furrowed material is seen to have a distribution like a flower's petals around the volcano. From these Viking pictures it seems at least possible that the furrowed material represents gigantic rockslides off the flanks of Olympus Mons. These slides are much bigger than any on Earth and have travelled for many hundreds of kilometres. If this explanation is accepted then the scarp would be scar remaining after the rockslides had taken place. Possibly, again, permafrost played a part in the production of these slides and if we assume that at some stage in the volcano's history, possibly during a period of long repose between eruptions, much of the outer flanks of the volcano became saturated with permafrost — possibly in ash layers — then a sudden rise in temperature resulting from renewed activity could cause melting and the resultant production of enormous landslides. This may seem a far-fetched idea and certainly we should not rule out the possibility that they indeed represent some form of volcanic activity.

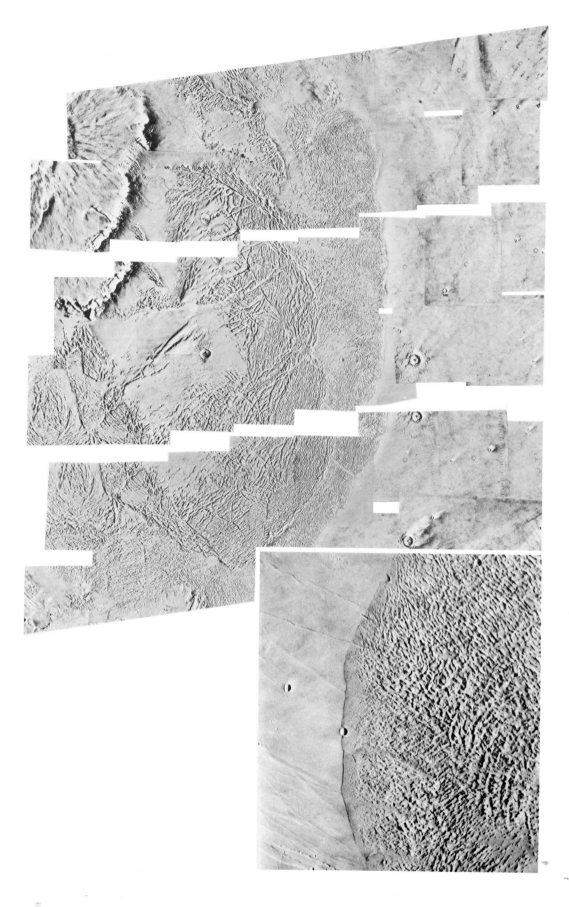

Mars: Tharsis Volcanoes

Olympus Mons is not the only large volcano on Mars. To the southeast of Olympus Mons is a volcanic region generally known as the Tharsis area in which there are three large volcanoes together with many smaller ones. Tharsis forms part of a large area of uplift, some 11km above the surrounding plains, and on top of this the three major volcanoes have their summits at about the same height as Olympus Mons. These three volcanoes are, from north to south, Ascraeus Mons, Pavonis Mons and Arsia Mons. The picture opposite shows in high resolution the caldera complex and surrounding flanks of Ascraeus Mons. A smaller caldera has cut part of the larger and older caldera; but later collapse in the larger caldera has then again cut the smaller one. Evidence of subsidence is given by the numerous fault scarps and graben that occur in concentric patterns around the inside edge of the caldera. The floors of the caldera have been flooded by later lava flows and details on the flows can be seen clearly. The inner walls of the calderas show clear evidence of later degradation giving rise to scree runs. Outside the caldera we see many hundreds of flow fields with lobate flow boundaries and numerous sinuous lava channels. Some of these can be traced back to vent areas on the upper flanks of the volcano near to the caldera margin. These consist of small pits associated with graben features and chains of pits running parallel to the edge of the caldera. It would appear that many lava flows on this volcano originate from these fractures that ring the summit of the mountain. This is an unusual situation, for on many volcanoes on Earth the eruptions are generally only associated with linear rift zones cutting across the volcano. Thus eruptions tend to come from fissures that are radial to the summit and are concentrated in specific areas of fracturing. However, there are some volcanoes on Earth that operate like this one on Mars; for example, some of the volcanoes in the Galapagos Islands and elsewhere in the Pacific have well defined fractures surrounding the caldera from which lavas have erupted.

Let us try and imagine what this volcano would look like if we were able to cut it like a cake. The lavas are clearly fed to the surface through fractures that form cylindrical rings around the volcano and, once an eruption has ended, it may be presumed that some of the lava will be left in these fractures to give cylindrical sheets of lava surrounding the central conduit. Is such a structure possible? One area where we can investigate the internal structure of a volcano is the 60-million-year-old volcanic area of Western Scotland. Here we see the eroded stumps of volcanoes where the top few kilometres have been removed. The characteristic structure of these eroded volcanoes is ring intrusions, suggesting that at least some volcanoes on Earth feed the surface lavas from ring fractures. Both the inside and outside of these large martian volcanoes are therefore comparable to the volcanoes that we see in some areas on Earth.

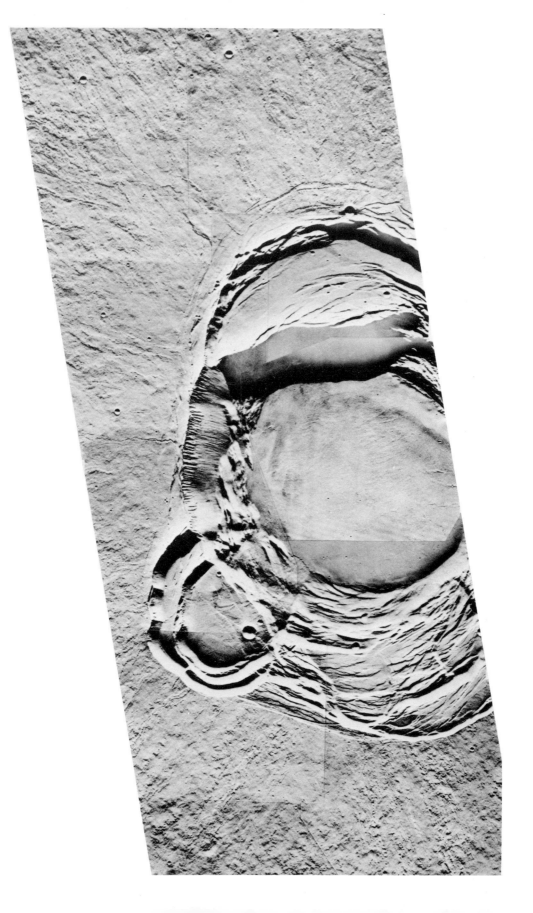

Mars: Arsia Mons

When Viking pictures were returned of the Arsia Mons volcano, geologists were particularly excited to notice on the flanks of the volcano evidence for an enormous landslide that originated not from the outer flanks, as we saw on Olympus Mons, but from near the summit. The picture shows this landslide. At the top of the picture we see the edge of the summit caldera, which is about 140km in diameter. Below this we see hummocky ground that grades into a series of fine concentric ridges. These surface characteristics are what we might expect from a large landslide, which in this case is some 600km long. There are also smaller landslides, as for example in the middle of the picture.

As we trace the landslide towards the summit caldera we find that the upper flanks of the volcano are scarred, and it is likely that in the later history of the volcano the upper flanks became unstable, possibly as a result of melting of permafrost in ash, and a large volume of material broke away and cascaded down the sides of the volcano to near the bottom of the slope.

Some scientists have suggested that, as the landslide came to rest, it exerted a strong outward pressure on the underlying ground causing the fine ridged features that we see running parallel to the edge of the slide. The ridges cut across an impact crater, demonstrating that the impact crater existed before the slide occurred; it is unlikely that the slide itself over-rode the crater, otherwise it would have partially buried it. The conclusion, therefore, is that the slide stopped short of this crater and that the edge of the slide is the edge of the hummocky ground. The ridges are not part of the slide but compression structures superimposed on the underlying ground including the impact craters. As we mentioned for Olympus Mons, such slides are not uncommon on volcanoes on the Earth, and large slides do occur on the outer flanks of volcanoes such as those in Hawaii; however, in Hawaii they occur under the sea, where presumably the presence of water helps to lubricate the slide. For slides on Mars it is feasible that they operated without the help of such lubrication, but the assistance of melted permafrost is clearly a strong possibility.

Although it seems most likely that this feature is a landslide, we cannot completely disregard the possibility that it is a form of ash flow or pyroclastic flow of volcanic origin emplaced from a density-current cloud consisting of a mixture of hot lava particles and gas; such an explanation must be considered in future investigations.

Before we leave Arsia Mons we should note the large graben that cut across the flanks of the volcanoes. These are most certainly related to collapse along the fault zones in response to tensional stresses, and it is this tensional environment that probably assists the uprise of magma from depth to the surface.

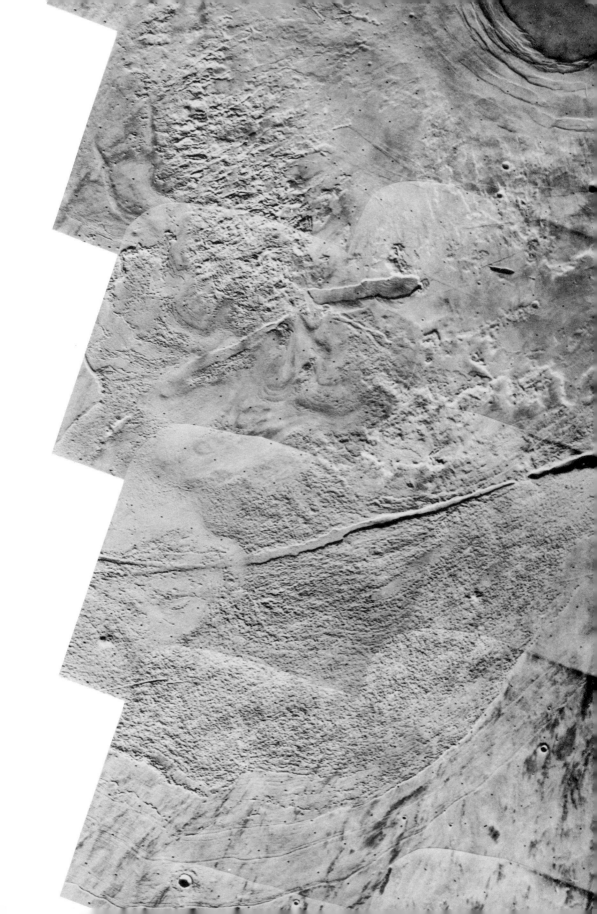

Mars: Alba Patera, A New Type of Volcano

The fifth major volcano on Mars is known as Alba Patera. Although this volcano is of large volume it is not so striking as the ones we have discussed previously because it is relatively low, having a height of only about 2km — although its diameter is some 1,600km. The form of the volcano is indicated by the name *Patera*, Latin for saucer, and indeed this volcano has the form of an upside-down saucer. Mariner 9 pictures of Alba gave the impression that the surface was relatively smooth but the better Viking pictures showed intricate details of lava flows. Thus the idea that Alba was an old degraded volcano was rejected on the Viking evidence, and we now have to consider Alba as a young volcano comparable to the volcanoes in the Tharsis region to the south of Alba.

The upper part of the volcano is cut by numerous graben running approximately north-south across the summit region. As they approach the summit, the graben diverge to form an incomplete ring around the summit region and within this ring we find a complex of calderas which are illustrated opposite (bottom). The older, larger caldera towards the top left appears to be largely filled by later lavas which have built up a small cone in the middle which in turn is cut by a caldera some 70km across.

The flows on Alba Patera are diverse in morphology. In the upper right picture we see examples of what have been called sheet flows, with well defined lobate lava fronts lacking in surface structure such as lava tubes, lava channels or festoons, and with relatively level surfaces. These are the most conspicuous type of flow on Alba Patera, and many appear to originate near the summit shown in the lower picture. They are very large, some being 300km long.

At the top left we see an area of flows that differs strongly from the sheet flows. Here we see long ridges, many of which have sinuous channels along their crest. These flows are considered to have been formed by lava moving in tube and channel systems.

Both of these types of flow and others are not unlike many flows seen on volcanoes on Earth, and the only major difference is that the features on Alba Patera are as much as ten times bigger than any such features on Earth.

The variety of types of feature suggests that lavas of different compositions and therefore different rheologies have been erupted by this volcano; but it is more likely that the conditions of eruption were different. The way in which a lava flows, once it reaches the surface, depends on many parameters, including rate of effusion, viscosity, yield strength, slope of the ground and gas content. Probably the most important of these is the rate of effusion, those lavas that are erupted at the highest effusion rate travelling long distances while those erupted slowly pile up close to the vent. However, factors such as gas content can affect considerably the state of the lava, those lavas in which the gas is dissolved being more fluid than those where it has escaped to froth up the lava. Examination of many volcanoes, such as Etna on Earth, where the composition of the lavas remains remarkably constant, shows that many different styles of lava flow result from different eruptions depending on variations in the factors mentioned above. Alba Patera, therefore, provides us with a volcano full of diversity and interest and will allow us to study more the effects of the physicochemical controls on volcanic activity.

Mars: Flood Lavas

We have seen on the Moon how the *maria* consist of extensive floods of lava of basaltic composition. Similar lavas occur on Mars. In the picture opposite we see the edge of a great sheet of lava that has flowed down from Arsia Mons in the Tharsis region. The lavas seen here have finally come to rest more than 1,500km away from the summit of that volcano and have flooded into depressions on the margin of the southern-hemisphere cratered terrain. Well defined flow fronts are seen within the large crater Pickering, which is 120km in diameter. From the form of the flow front it can be seen that the lava was forced to flow around the high area in the middle of the crater formed by the central peak.

These flows that extended so far away from the vent area must have been very fluid and may well be similar in composition to the basalts of the Moon. Nevertheless, it was probably again the conditions of eruption that allowed such extensive flows, and based on our understanding of physical controls of lava-field form we may infer that such flows were erupted at very high rates, of the order of thousands of cubic metres per second, from fissures in the vent areas. With these high rates of effusion lava would be forced to travel for long distances owing to the high hydrostatic head built up over the vent area, the lava ponding to great depths over the vent and, while still retaining its fluidity, flowing away at a high rate. Flows of the type seen opposite cover very large areas surrounding the major volcanoes, giving a radial extent to each of the major volcanoes of more than 1,000km.

Mars: Minor Volcanic Edifices

Not all the volcanoes of Mars are so incomprehensibly large. In the Tharsis region and also in Elysium there are volcanic constructs of more nearly the size that we are familiar with on Earth. In the pictures opposite we see several of these volcanoes, in the Tharsis region. In the Viking picture mosaic (bottom) we see smaller volcanoes on the northeast margin of the Tharsis ridge. The volcanoes are associated with an area of intense faulting, producing the long graben-like features. The larger of the two features shown here is Ceraunius Tholus, which is about 120km across. This volcano has a summit caldera, but at the bottom of the flank there is an elliptical crater with a 'butterfly wing' distribution of ejecta around it typical of oblique impacts. Running from near the summit caldera down into the impact crater is a well defined channel, and one interpretation of this feature is that the impact itself triggered the volcanic activity which caused it.

The other two volcanoes illustrated (top) are also small: the one on the left is the 50km-diameter Biblis Patera and on the right is the similar-sized Ulysses Patera. Both these volcanoes are cut by graben and, in the case of Ulysses, there are two superimposed impact craters, the ejecta of which is seen on the floor of the caldera.

It will be noted that both Biblis and Ulysses are completely surrounded by younger lavas, which presumably bury a more extensive base to the volcano. The original height and size of these volcanoes is therefore unknown, and they are probably considerably bigger than might be suggested by the parts that now stand above the surrounding lavas. However, it is unlikely that these volcanoes before they were buried were as big as the Tharsis giants, such as Arsia Mons, because their summit calderas are somewhat smaller.

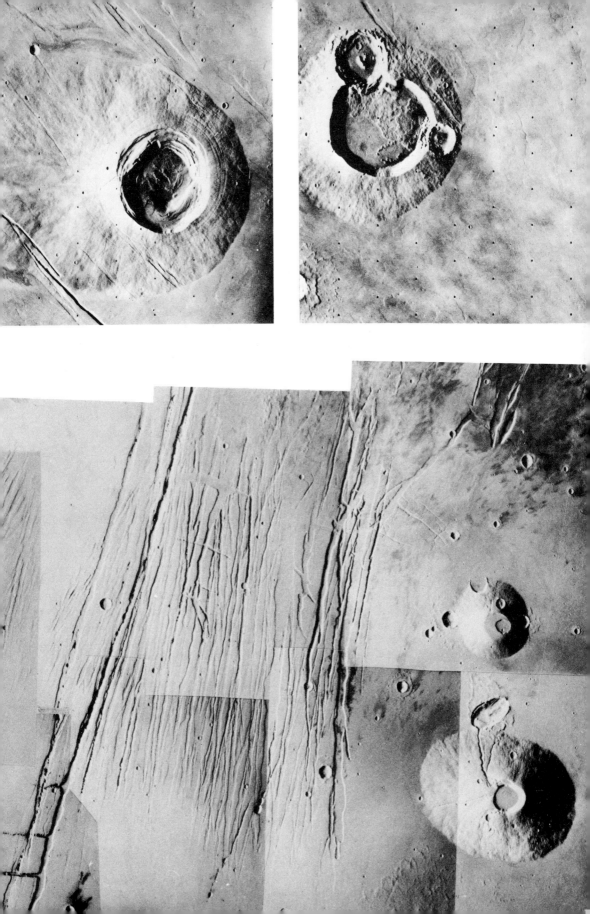

Mars: Old Volcanoes

The martian volcanoes we have looked at so far in this book have all been relatively fresh; however, there are many examples on Mars of volcanoes that are degraded. One example of a volcanic region that is relatively old but still recognizably volcanic occurs near to the Hellas basin in the southern hemisphere. Here we see volcanic plains and occasional prominent volcanoes such as Tyrrhena Patera, illustrated opposite. Although this volcano has been dissected by channels its low conical form is still visible. The summit is marked by an irregular breached collapse surrounded by concentric graben fractures. Dissection has been sufficient to remove any surface textures of the volcanic materials forming the flanks, and it is thus not possible to determine whether the volcano is made up of pyroclastic deposits or lava flows, let alone what kind of lavas they might be. If the volcano is made up of lavas then later sustained torrents of water are required to produce the observed dissection. However, pyroclastic rocks would erode more readily. Since the volcano has a low profile it is possible that it is constructed of ash-flow material formed by dense clouds of lava particles and gas. Such ash flows on Earth may travel for hundreds of kilometres emplacing ash deposits that may be hundreds of metres thick. That the volcano is made up of less resistant pyroclastic material is also suggested by the mesas that form outliers in the surrounding terrain.

Large extents of flood lavas appear to occur in the surrounding region, and some of these may be younger than Tyrrhena Patera. A volcanic origin for this surrounding material is suggested by the ridges similar in appearance to the *mare* ridges on the Moon. Some of these are seen near the edges of the photograph opposite.

Tyrrhena Patera is just one example of older volcanic terrain. In the northern hemisphere plains there are many areas where lobate fronts are seen, indicative of sheets of lava that have spread over the plains burying the older topography. There is still debate about the plains that lie between the craters of the southern hemisphere, as these also may have been formed by volcanic effusions early in the history of Mars.

A study of martian volcanism is clearly of extreme importance in understanding the history and internal constitution of the planet. The rocks formed by volcanic eruptions are, apart from those thrown out by impact cratering, the only rocks that have come from deep within the planet. Magma, if it is to rise, must have a lower density than the surrounding rocks, and is pushed up by the weight of rocks around it. The maximum height to which a volcano may be built therefore depends, at least in part, on the depth of melting. Thus the height of the volcanoes gives us some indication of the thickness of the lithosphere, which on Mars must have been considerably thicker than the Earth's lithosphere at the time when the volcanoes of the Tharsis region were formed.

Geological experience shows that volcanic activity is the principal crust-forming process and it is likely that much of the old crust of Mars, now covered by large impact craters, was once volcanic terrain that has now been battered by impacts reducing the surface to layers of brecciated debris and leaving few relics of the original volcanic forms.

Mars: Graben

Mars has had a much more active crustal history than the Moon or Mercury. Many more fractures are visible and the numerous volcanoes of the planet are evidence that materials of the interior have used these planes of weakness as a means of reaching the surface.

Opposite are two pictures, both of areas about 50km from top to bottom, showing examples of the most conspicuous landform to result from faulting, the graben. As on the Moon, this consists of a long straight narrow block of crust dropped by what may be several hundred metres relative to the crust on each side, but on Mars graben occur in such dense swarms that they frequently overlap and intersect one other, giving the complex appearance opposite. Several sets of fractures with different orientations are visible in the lower picture. Many of the crater-like and elliptical depressions in both images (and certainly the large one at A, and most of those arranged in chains) are not impact features. A large number are collapse depressions, while others may have been volcanic vents similar to the lunar ones on the Hyginus rille (page 37). It is interesting that the linear chains of depressions do not trend in the same direction as the graben.

The crust of Mars includes two striking asymmetries. Firstly there is the dichotomy between the northern low plains and the southern heavily cratered uplands described earlier. The second hemispheric asymmetry consists of a very densely radially fractured hemisphere around a large crustal bulge (under the great Tharsis volcanoes, sometimes called the Syria Rise), opposite what appears on first examination to be an almost unfractured hemisphere. No overwhelmingly convincing explanation has been put forward for either of these great contrasts. It does seem that the division of the planet into a low northern and high southern hemisphere occurred before the crustal doming responsible for both the Tharsis volcanoes and the radial faulting.

Why the northern of these two 'geologic hemispheres' should have apparently been depressed several kilometres with respect to the other hemisphere is one of the greatest of the planet's mysteries. Some evidence of the sequence of events has been preserved along the boundary. Most striking is the fretted terrain which commonly occurs here and which appears to have developed through the erosion of the southern cratered hemisphere materials. It is very difficult to see how or why the pre-existing ancient cratered surface of the northern hemisphere could all have been consumed in this way. The existence of isolated outcrops of old surface within the low northern plains suggests that most of it has merely been downdropped along faults and then buried by young materials.

The subsidence of northern crust has usually been explained in terms of pervasive fracturing of the planetary surface by some internal expansion. Internal expansion is quite easy to account for — a mineral phase change giving a volume increase at depth in response to rising temperatures, for instance. It is not so easy to explain why fractures should have had so much more drastic consequences in the north than in the south. Perhaps the crust was thinner there, but why?

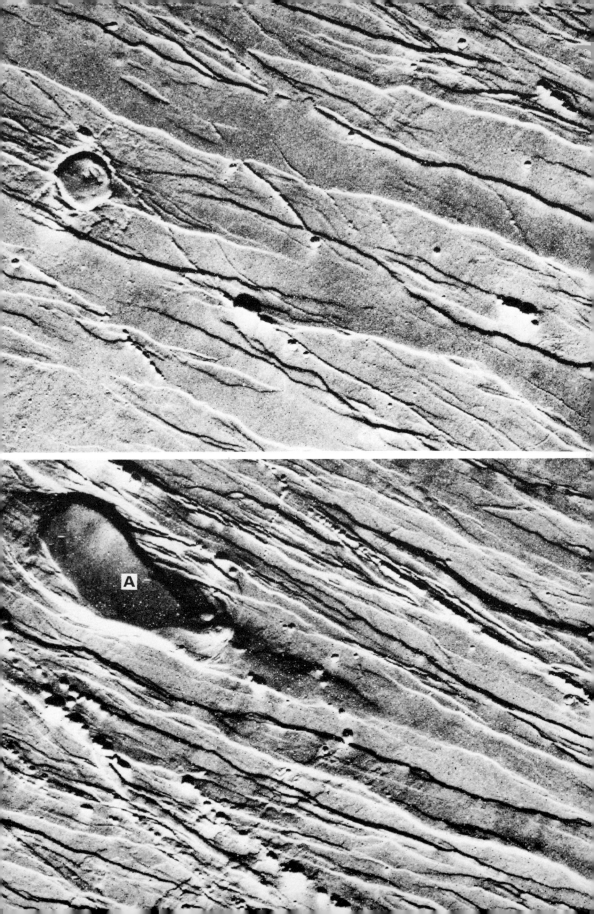

Mars: Syria Faulting

The illustration opposite shows a large area, about 1,500km from top to bottom, which contains some of the most intense collections of faults around the Syria Rise, including part of the Claritas Fossae, at C.

Though faults can be found with almost all possible orientations around Syria, as Mike Carr of the US Geological Survey has commented, two sets are particularly prominent: those running roughly north-south and those running northeast-southwest. The Claritas Fossae form the south-trending arm of the former. The northern arm is represented by an intense fault set around the old volcano Alba Patera. The Alba faults are bent gently around each side of the volcano, implying that they formed after it. To the east of Syria stretch the great canyons of the Valles Marineris. The fact that they too are radial to the Syria Rise suggests that at one time they might also have been merely graben. They have since been enormously enlarged and altered. The Valles Marineris are considered separately on pages 146-55.

The deformational history of this area has been long and complex, but the observed fracture pattern remains generally consistent with the stress distribution which would have resulted from regional doming. Looking closely at particular areas, several sets of graben with slightly different orientations can be recognized. Also, as can be seen opposite, where younger rock units have buried old fractures, these have often subsequently been lightly faulted.

The detailed order of events in the fractured terrain has proved difficult to work out. Four main episodes of activity following early cratering have been proposed: the development of the volcano Alba Patera; northeast to southwest fracturing; more volcanism (the great Tharsis ridge volcanoes); and finally north-south faulting. Whatever the details, it is clear that the Syria uplift was initiated very early in post-accretional Mars history, long before the present volcanic cap was emplaced. The several episodes of faulting correspond to continuing, possibly episodic, vertical motions.

Though we are far from understanding martian crustal bulging, several mechanisms have been proposed. Perhaps a deep upwelling in the mantle, driven by convection, was responsible. Perhaps chemical differentiation produced changes in Mars' dynamical properties and led to equatorial buckling.

Close examination of the picture opposite reveals some of the complexity of the Claritas Fossae. Its faults dominate a sequence of largely volcanic rocks. The radially textured massifs at A and B may be much eroded volcanoes. Many possible smaller and younger volcanic hills and domes are visible, particularly on crater floors.

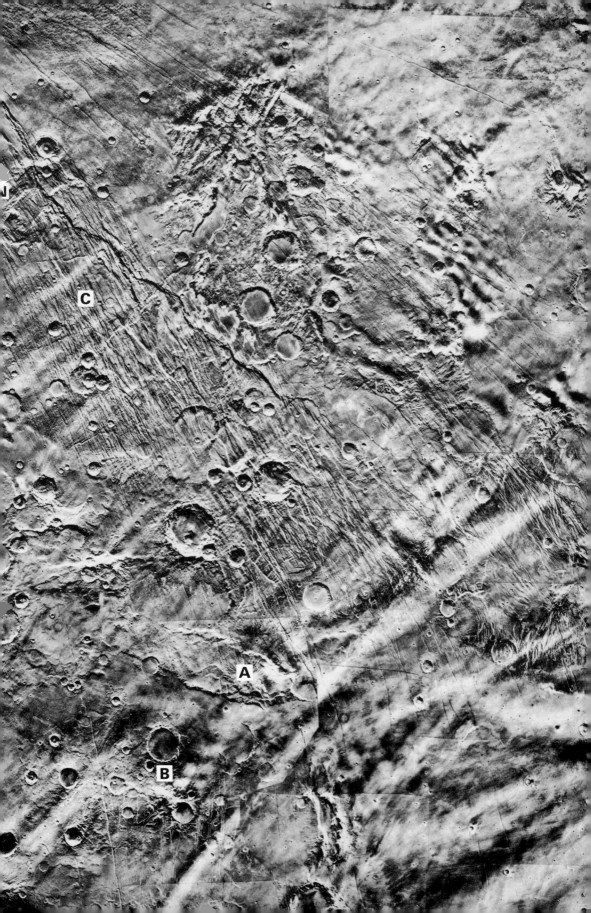

Mars: Other Fractures

The graben shown on the last two pages are only the most prominent of very many narrow depressions of probable tectonic origin. The lower right inset opposite, for instance, showing an area 60km across, illustrates on the left a fairly typical set of graben but on the right a complicated rectilinear pattern. The rectilinear pattern is made up of many interconnecting troughs of various widths and orientations. Most trough sections are linear, although some are gently arcuate. Several low ridges are cut by the troughs, which must therefore be younger than them. It is possible that the ridges are volcanic dykes and that the whole area has been massively intruded at depth. The rectilinear fractures may be the response of the surface to the stresses of intrusion.

The upper left inset, 80km from top to bottom, includes troughs of extreme sinuosity which have been variously explained as dry water channels or as being analogous to the sinuous cracks of paving stones. There can be little doubt that the deepest channels in the image have been water cut. A smoothly modulated width, the teardrop mid-stream islands, the correspondence of the apparent flow direction to the regional slope—all support this idea. The much more sinuous depressions on both banks are more difficult to account for. The main line of evidence for a non-river erosion origin for the narrow channels lies in their discontinuity. They also have many straight sections, and this implies possible fault control. Downstream directions obtained from trough widening and from junctions between troughs are in many cases opposed to the regional slope (although they may correspond to more local slopes). Some of these features may be pressure release fractures along which material has slipped towards the channel, long after the end of water flow, but most appear to be cut by (and therefore predate) the main channel.

The main picture opposite, which shows an area 50km across, is located in the northern midlatitudes of Mars, in the region originally earmarked for the second Viking landing. It is dominated by an irregular system of cracks, each a kilometre or so in width. Here there is no obvious pattern to the arrangement, but nearby crack rings and crack polygons are common. Erosion by no conceivable fluid (air, water, lava, or something more exotic) could have produced such a complex discontinuous arrangement of valleys. There is general agreement that we are looking at the consequences of tectonic activity; that is, of ground motion without much erosion. Argument continues, however, on the driving force of the motion. The contraction through cooling of massive flood lavas has its advocates. The lava flows would have had to have been very thick, though—perhaps hundreds of metres—and emplacement would have had to have been very rapid. An alternative explanation, which has often been invoked but rarely worked out in detail, is that the surface has been moulded by ground ice activity. In the frozen areas of the Earth so called 'patterned' ground, with polygonal cracks and mounds, has frequently resulted from motions generated by freeze-thaw, heating-cooling, and wetting-drying cycles. On Mars we again have a problem of scale. On the Earth, patterned ground is dominated by features a few metres or tens of metres in diameter; martian crack rings can be tens of kilometres across.

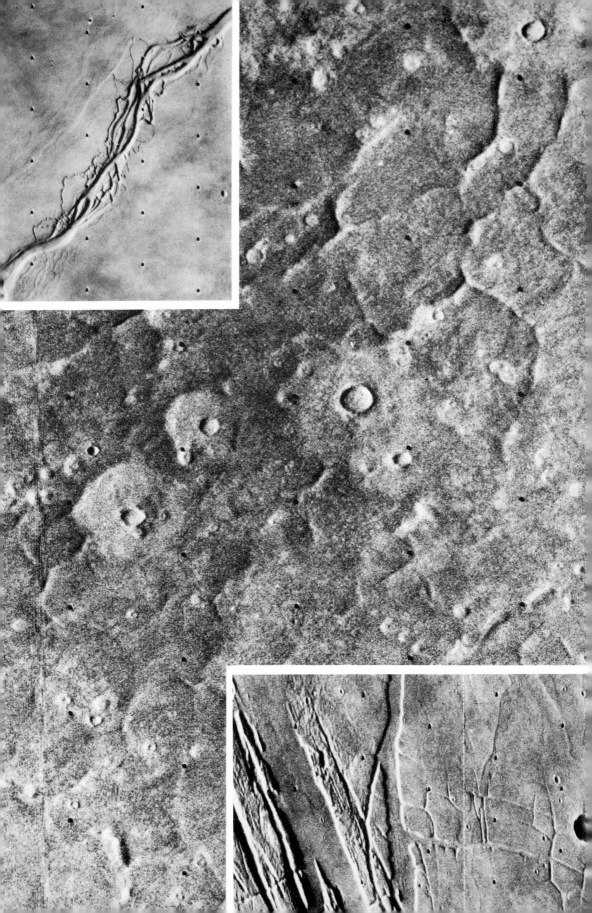

Mars: The Valles Marineris

Volcanoes, craters, perhaps even river valleys might have been expected. Wholly unexpected was the discovery by Mariner 9 of the vast and complicated canyonlands of the equatorial regions of Mars. The picture opposite shows the western part of the canyonlands and covers an area more than 800km from top to bottom. Named the Valles Marineris in the spacecraft's honour, they are about 4,000km long, and are more than 700km across in places, and reach depths of 6km. Some of the component canyons have a startling resemblance to the Grand Canyon of Arizona, but together they dwarf it in scale. Duplicated in North America, the Valles Marineris would stretch from California to New York.

The Valles Marineris run west to east from the summit of the great crustal bulge under Syria Planum to the source region of the Chryse channels. Central troughs of the canyonlands lie along the crest of a shallow crustal ridge. Going from west to east the nature of the system changes greatly and our discussion will recognize several provinces. The change from one province to another is in all cases gradual, and it is clear that the Valles Marineris form an inte-grated system produced at the same time, subject to many of the same modifying processes. Closest to the summit of the Syria Rise lies the Noctis Labyrinthus (shown over-leaf). This tangle of collapse depressions and graben seems to have been the least modified by subsequent erosion. Stretching away for 2,500km to the east lies the linearly troughed terrain forming the canyonlands proper. At their eastern extremity these troughs grade into the chaotically jumbled terrain whose decay seems to have fed the several northward-flowing channels of Chryse with their water.

If we could only understand the formation of the Valles Marineris system, we would be a long way further in our understanding of Mars. First and foremost, the canyonlands form one of the principal martian lineaments. Several of the principal lineaments have simple geometric relationships with each other. Running at 90° into the Valles Marineris, from south of Labyrinthus Noctis, are the Claritis Fossae, which as we have seen is the planet's zone of greatest faulting. Running across the junction of these great structures, making an angle of 45° with both, is the Tharsis Ridge, the planet's most important volcanic lineament. Surely this simple geometry must reflect fundamentally important crustal relationships. Observation of com-parably simple geometries on the Earth led to the powerful explanatory theories of sea-floor spreading and continental drift. It is natural to ask if Mars has a small number of mobile crustal plates like the Earth. Does the Valles Marineris, for instance, represent a rift where two plates have moved apart?

Mars: Canyonlands

Labyrinthus Noctis, illustrated opposite, consists of an interlocking web of elongated collapse depressions, crossed by a fine network of long, shallow graben. It adjoins the Valles Marineris system at the top left of the picture, which overlaps with the bottom of the picture on the previous page. The elongate depressions of the central region resemble the 'turtle structures' common on the summit regions of large terrestrial domes. The picture includes almost the whole of the Labyrinthus Noctis area. The broad volcanic plain at S is Syria Planum while the feature at H received the informal name of 'the Chandelier' during the Mariner 9 mission.

The troughs which dominate the central region of the Valles Marineris (previous page) are generally linear, steep-walled valleys, several hundred km long and several tens of km wide. They are arranged in parallel groups with plateaux of similar dimensions in between. Some troughs are connected, while others either pinch out or finish abruptly.

Three groups of canyons may be recognized from east to west. Joined to the Labyrinthus Noctis are two long, narrow canyons: Tithonium Chasma in the north and Ius Chasma in the south. These western canyons have generally steep walls, varied by huge slump features and a number of tributary canyons. Further east the Valles Marineris attain their maximum width, with four main canyons running parallel.

One of these east central troughs, Hebes, is a completely closed depression. The other three, Orphir, Candor, and Melas, have been so broadened as to join with each other. To the northwest, a smaller lower canyon, Echus Chasma, which may be partially choked with sediment, opens out to the northern plains and connects with the fluvial valley Kasei Vallis. Further east still the system narrows again and simplifies to a single, strikingly linear trough, Coprates Chasma.

Beyond Coprates Chasma lies a vast area of mixed canyons, chaotic terrain and river valleys. The canyons (Ganges, Capri and Eos Chasma) are lower than elsewhere in the Valles Marineris, have less linear walls, and trend generally northeast-southwest rather than east-west. They constitute a transition from the troughed terrain to the west to the chaos and sediments of the Chryse area to the northeast. The area around Capri Chasma was selected before the Viking mission as a possible site for the second landing. A landing near the edge of a canyon would have enabled both the study of an area of old martian highland (a complement to the first lander's observations of young lowland) and perhaps sampling of wind-transported canyon materials. In the event, as we shall see, the second lander was placed in Utopia Planitia, a safer, flatter area and one favoured by the biologists among Viking scientists.

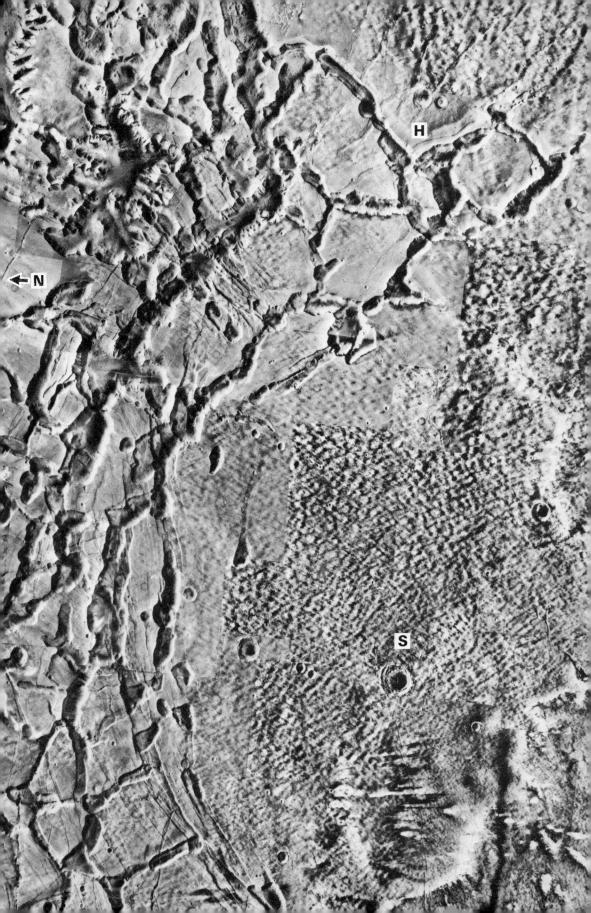

Mars: Canyon Walls

Several distinct types of canyon wall have been observed. Some of these can be seen in the picture opposite, which shows an area of Ius Chasma 50km across. Among the most common are spurs and gullies, landslide alcoves and tributary canyons.

Spur and gulley walls consist of alternating rocky spines and short blunt valleys, as on the left of the picture opposite. The spurs most frequently run downslope, but in some areas other orientations are also seen, suggesting structural control by joints or faults. Martian spurs appear to have developed in great uniform thicknesses of rocks (without noticeable variations in resistance to erosion with height). This is consistent with the observation that spur and gulley morphology on Earth is dominant in moderately resistant, uniform rock sequences.

Tributary canyons have a more restricted geographical distribution. They are most prominent on the southern side of Ius Chasma (right in the picture opposite), where they characteristically have blunt heads and pronounced V-shaped profiles, with little or no flat floor. As B. Lucchitta of the US Geological Survey has reported, in some places where tributary canyons have their own tributaries, these do not share the same floor level as the main valleys. Such 'hanging' valleys are common on Earth. Their presence on Mars indicates that, while erosion ceased in some tributaries, it often continued in others. Perhaps tributary canyon growth was very episodic and irregular. As with spurs and gullies, the linear elements of tributary canyons do not seem to be randomly orientated and structural control of erosion is again implied. The concentration of tributary canyons on the southern side of Ius Chasma, rather than the northern, has been explained in terms of the regional slope of the area. On the Earth, the regional slope into or out of canyons has been observed to determine the rate of their development. They seem to erode much more rapidly up a slope than down one, just the situation on the south side of Ius Chasma.

Canyon wall type does not seem to alternate in a systematic way throughout the Valles Marineris. An interesting observation is that no gullies or tributary canyons seem to have developed on the walls of landslide alcoves. Either landslides are generally more recent than the other landforms or old landslide scars are soon obscured.

Many canyon walls reveal clear signs of layering in their rocks. Layering is particularly apparent in the upper third of wall slopes. Light and dark bands and breaks in slope argue that the materials into which the canyonlands have developed are vast successions of quite thin flat-lying rock units. Important features of canyons, detected for the first time by Viking, and first described by Karl Blasius of the Planetary Science Institute and his co-workers, are low straight scarps, common at the foot of canyon walls. These often cut across other wall details, such as spurs and gullies, leaving them hanging. Basal scarps are clearly young features and only a few land-slides seem to postdate them. They appear to represent major fault traces, along which recent canyon deepening motions have occurred.

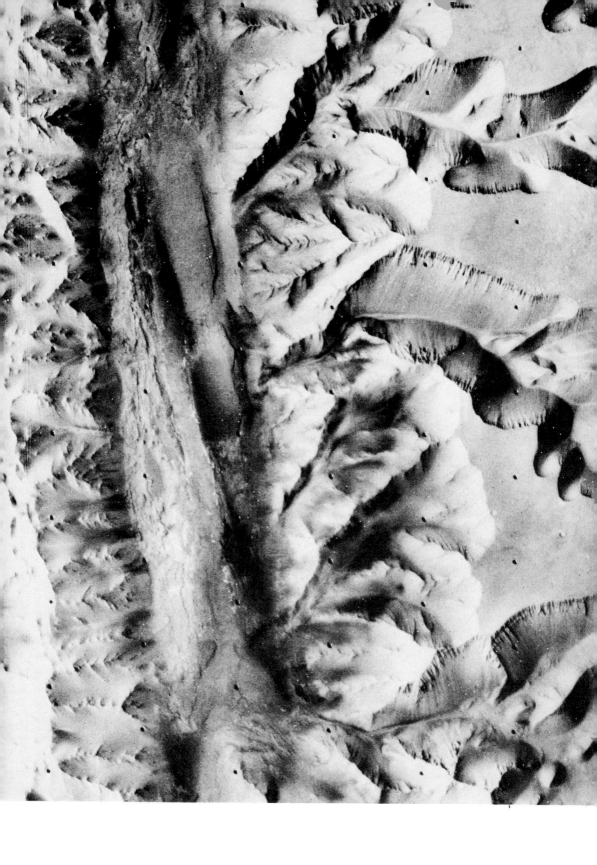

Mars: Canyon Floors

Canyon floors are generally very rough when contrasted with the plateau surfaces around them. They are also remarkably varied, with several kinds of plains surfaces and many hills, plateaux and ridges. Floor plains with surfaces similar to the smooth and cratered neighbouring uplands are widespread. They are most conspicuous within the Labyrinthus Noctis and the westernmost canyons. Many are obviously instances of downdropping, within graben, of otherwise barely disturbed highland blocks. Probably such down-dropping has occurred throughout the Valles Marineris, but elsewhere the floors so created have been subject to much deposition and erosion.

Perhaps the most spectacular floor units are those composed of great landslides. A number of these are visible in the picture opposite, emanating from both sides of the canyon wall. The largest of them is 100km long. Lobate landslide deposits mantle large floor areas. Often there is a blocky inner lobe surrounded by a smoother, broad outer lobe. Inner lobes typically are crossed by ridge and valley systems, roughly parallel to the source canyon wall (at right angles to the slide direction), while outer lobes are crossed by fine striations along the probable slide directions. The ridge and valley systems can be explained as the result of buckling of the slide as the forward parts decelerated. Fine striation patterns similar to those of Mars have long been known on terrestrial flows, but have not been satisfactorily explained. Both the landslide types just described can be clearly traced back to landslide alcoves from which the slide materials collapsed. A number of narrow, thin slides can also be seen at the mouths of tributary canyons and large gullies, and even smaller features with leveed (raised) rims have been tentatively interpreted as mud flows.

In the eastern canyons a large area of floor is occupied by closely spaced, often conical hills. These hills, in regions transitional to chaotic terrain, are remarkable for their uniform slopes.

Small areas of fractured floor are widely scattered. They resemble on a small scale areas like the Labyrinthus Noctis and almost certainly have resulted from renewed vertical motions of downdropped blocks. In places fractured floors are associated with the canyon wall basal scarps, and both features seem to imply comparatively recent motions.

Striking features of many canyons are interior ridges and plateaus. Many of these were observed from Mariner 9 imagery to be layered, evidence for which included examples of both apparent differential erosion and of albedo striping. In places up to thirty layers of quite uniform thickness can be traced in Viking pictures. Cyclical sedimentation seems to be implied, perhaps driven by periodic climate changes during formation. Two hypotheses have been proposed. One argues that the layering is widespread throughout the martian materials into which canyons have developed. The other suggests layered deposition after canyon formation, followed by renewed erosion. The former is generally the more plausible hypothesis although evidence exists (in Candor Chasma, for instance) that the latter mechanism has operated locally.

Mars: Chaotic Terrain

Although nowhere else on Mars has troughs as impressive as those of the Valles Marineris, canyons are present elsewhere, notably near the volcanoes Elysium Mons and Hadriaca Patera. On Mars canyon formation seems to have been intimately linked with volcanic construction.

Related to true canyonlands are the chaotic areas south of Chryse Planitia and joined to the Valles Marineris in the west. A typical area of chaotic terrain, 700km from top to bottom, is shown opposite. These complex terrains, consisting of jumbles of blocks filling irregular depressions, are found within an area 2,000km square centred on 30° west, 10° south. Some occurrences are within closed depressions, while others are continuous with large ancient river channels. In plan view, chaotic terrain is very irregular, in contrast to the often quite simple and symmetric troughs of the canyonlands to the west.

We are still not able to propose a convincing history of the origin and development of canyons and associated features, although some parts of the story have been much clearer since Viking. The Valles Marineris system was certainly initiated by extensional faulting radial to the centre of the Syria Rise. Downdropping of large blocks between faults to produce graben followed with continuing extension. In other places, canyons seem to have grown more from the enlargement and coalescence of small pits, rather than from long, straight-walled graben. Similar pits are common in the Earth's volcanic regions where they indicate collapse after the withdrawal of magma. Martian pits, though, do not appear to have been associated with local volcanism and their origin remains a mystery. Whatever pits are, pitting, like the motions along basal wall faults, may remain an intermittently active process. Certainly, it now appears that much of the volume of the Valles Marineris canyons has been created by downward motion.

The considerable differences between true canyonlands and chaotic terrain have been explained in terms of different processes of formation and more simply in terms of different geological setting. It is possible to argue that, in chaotic terrain, volcanic and tectonic activity merely acted as a trigger for ground ice melting, which precipitated collapse and carved channels. Alternatively, ground ice melting can be seen as a comparatively minor contribution to collapse, with graben formation being as important as in the canyonlands. The different morphologies can then be explained by different thicknesses of different rock types and by contrasting types of faulting.

Even canyon wall modifying processes seem to have been controlled by tectonic setting. Tributary canyons seem to have grown where facilitated by upland faulting. The most massive landslides have occurred where the local upland is crossed by graben parallel to the back wall of the resulting landslide alcove.

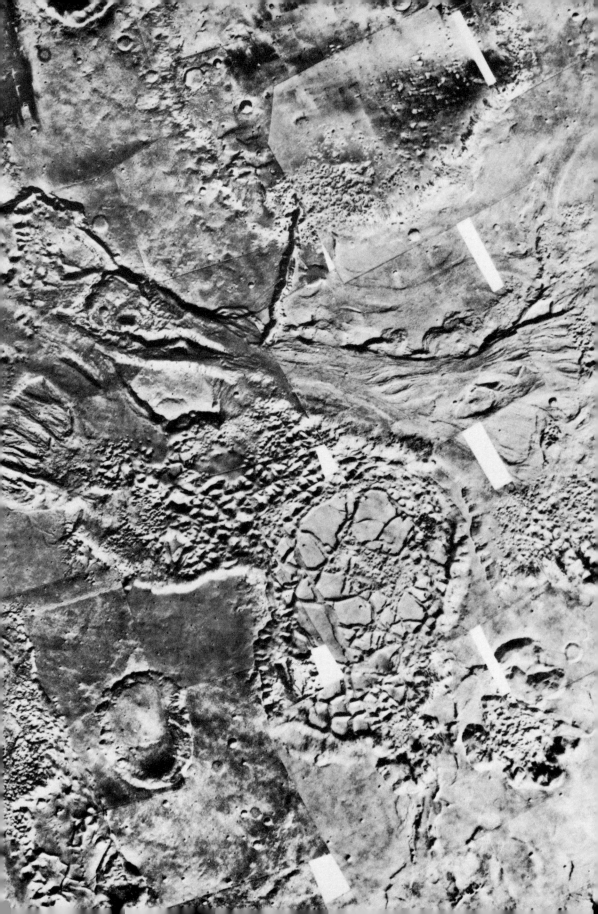

Mars: Water Channels

There has been speculation on the subject of rivers on Mars for more than a century — a considerably longer period than there has been any evidence for their having ever existed! A hundred years ago several observers, most prominently Percival Lowell, straining beyond the resolution limits of their telescopes, saw the narrow dark markings which came to be known and interpreted as canals, as described earlier. Astronomers were soon vying with each other in the complexity of their 'canal' maps and the extravagance of their theories. The favourite explanation of canals was that they constituted an artificial irrigation system, which was the work of intelligent Martians.

Even at the height of 'canal mania', many people refused to be intimidated into seeing what wasn't there. By the time of the first spacecraft mission to Mars, a large consensus had correctly dismissed canals as an optical illusion. The small number of low-resolution pictures returned by Mariners 4, 6 and 7 showed not only no signs of rivers, past or present, but together with the other data returned indicated a martian surface almost unaltered since the planet formed, and an atmosphere extremely hostile to running water.

A piece of dissenting evidence was the detection by Mariners 6 and 7 of areas of collapse, one explanation of which was that here volumes of frozen ground had thawed releasing water. Mariner 9 provided a spectacular confirmation of this idea. Vast areas of apparent permafrost decay were seen together with a whole variety of channels, discussed first by Dan Milton of the US Geological Survey. Enormous channels connect collapsed areas to the northern plains. Many smaller features have systems of tributary channels, and in the southern hemisphere small closely spaced furrows are visible on many slopes.

The channel system opposite, 250km across, is one of the most mature so far documented with five large tributaries (A, B, C, D and E), running eventually into the main channel F. Note that this system appears to be older than most of the bowl-shaped craters of the area (X and Y, for example) but younger than many of those with flat floors (the tributary E, for example, has clearly entered crater Z, although more recent crater-floor materials have buried the crater interior section of the channel).

The problem now is to explain how all these apparent dry river beds could have been produced when flowing water is quite impossible under the present atmospheric conditions of Mars. Not only is water not stable on the surface, there is little precipitable moisture in the atmosphere. There must have been a very different climate in the past to allow water erosion, and much more available water.

Mars: Climatic Change

Climatic changes are most easily produced by temperature changes, and several mechanisms are available to produce these over long periods (millions of years), and have been considered at length, notably by Carl Sagan at Cornell University. The two which have had most attention have been periodic planetary orbit changes and variability in the Sun's energy output. Planetary orbits are continually changing by small amounts in response to perturbations, particularly those of neighbouring planets. Over long periods these changes considerably affect the amount of solar energy received by different parts of the planet. The Sun itself may emit different amounts of energy at different times. We need to answer the following questions: How much water is around at the moment near the martian surface, and how is it distributed? How would conceivable temperature changes affect the present equilibrium? Are the dry river beds the result of one wet period, several, or many?

It is quite easy to see how there was enough water at some time in the past to erode the observed features. All the gaseous molecules trapped in the planet as it formed out of the solar nebula will have diffused more or less rapidly towards the surface. Carbon dioxide has outgassed to a considerable degree, and now forms most of the atmosphere and much of the polar caps.

Water should have outgassed at a similar rate, but its present whereabouts is difficult to determine. Depending on the model chosen for initial planetary composition, up to 1km of water over the whole surface needs to be accounted for. Most of this would have been outgassed soon after accretion, while perhaps 20m or so may have been contributed by recent volcanic activity. Possible 'sinks' for all this water include the polar caps, absorption on the surface regolith, ground ice, and escape into space. The polar caps may include up to a metre of water in their interiors. A very deep regolith could account for perhaps 10m or so of water. Unless atmospheric escape has been very much more effective than expected, ground ice must be the major sink for water. Corroborating evidence for this is abundant, in the form of the large areas of collapse which apparently fed ancient rivers with their water.

The channel illustrated opposite (Kasei Vallis) is one such collapse-fed valley. Its width remains quite constant for a great distance, indicating that it was carrying almost as much water at its 'head' as near its 'mouth'. Its few tributaries are discordant with the main channel and may be more recent features. In contrast, other channels progressively increase in width from 'head' to 'mouth' and have more integrated concordant tributary systems.

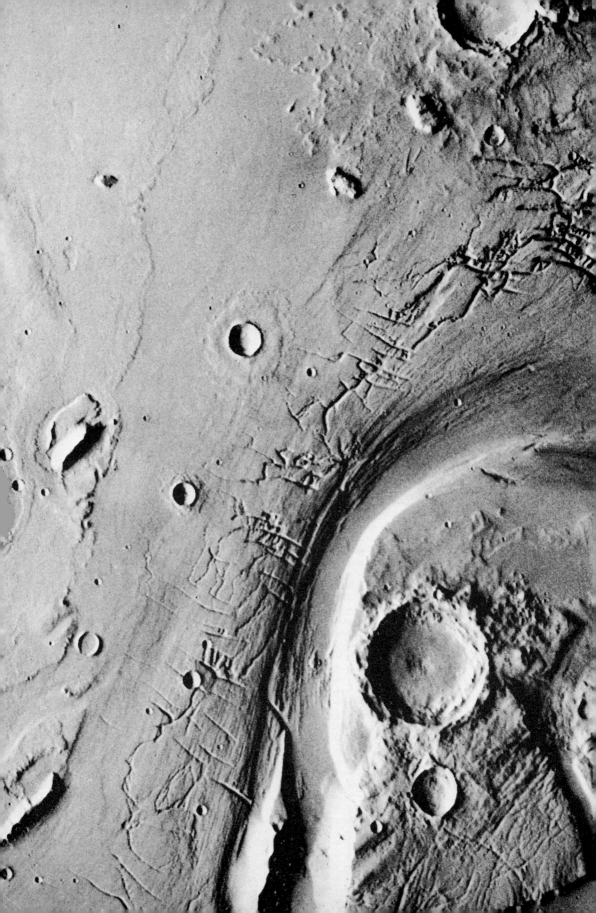

Mars: The Variety of Channels

As we have seen, Mars has several kinds of channels. There are the great broad features of the equatorial regions, numerous tributary systems in the southern hemisphere, and uncounted fine gulleys on crater rims and other slopes. The largest channels, which provide the best evidence for water action, are confined to four areas —south of Amazonis Planitia (Mangala, Ma'adim and Al Quahira Valles); between the chaotic terrain east of the Valles Marineris and Chryse Planitia (Simud Shalbatana, Ares and Tiu Valles); north of Lunae Planum (Kasei Vallis); and in fretted terrain (Auqakuh and Huo Hsing Valles). It can be no coincidence that all these areas lie along the great divide between the northern plains and the cratered highlands. Most of these channels can be traced back to areas of massive ground decay and appear to result from catastrophic permafrost melting and breakout. It is easier to explain the melting (by climatic change or by geothermal heating) than it is to explain why the water should have been so suddenly released. The rates of discharge of the observed channels seem to have been so great that water provided by the collapse of chaotic terrain can have supplied them for only a few days. The Earth has a few examples of such catastrophic breakout, when ice dams have burst.

Immediately after the Mariner 9 Mission, those martian channels with tributary systems were held to constitute the best evidence for rainfall once having occurred on the planet. Rain on Mars is much more difficult to achieve than brief surface flow. It would have required a dense atmosphere which would have needed much more carbon dioxide and water to be vaporized than has been positively located near the surface of Mars. Before Viking, the evidence was reexamined and the analogy between some of the martian systems and mature terrestrial river systems was questioned. The large-scale tributary patterns on Earth seemed to be missing on Mars. Tracing a terrestrial river channel from its mouth to source, one would typically pass a large number of junctions into more or less equal-sized tributaries. On Mars there is usually one main channel into which several or many tributaries run. martian tributaries also tend to join their main channels at more acute angles than is common on Earth, further evidence of immaturity or rapid formation. The channels shown opposite, in an area 140km by 100km, west of Chryse, include many such acute-angled junctions. Though Viking has confirmed the suspicion that many of the martian tributary systems initially attributed to rainfall may instead be shallow groundwater-fed features, a minority may be described as dendritic and a number of deeply incised examples may still support a rain-fed origin.

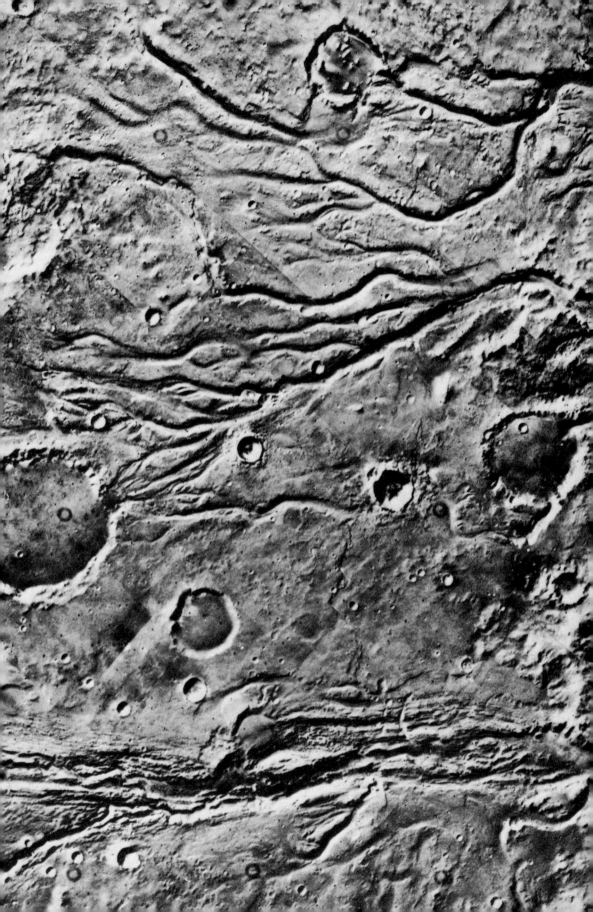

The furrowed terrain of the south equatorial region also seems to require rainfall to have occurred in the past. Two examples of furrowed terrain are illustrated opposite. The upper frame includes not only furrows (at F) but an area of partially buried fretted terrain at T (see overleaf for a discussion of such terrain). The narrow rim crests of craters appear incapable of providing, through ground ice decay, sufficient water to have eroded furrows.

Furrowed terrain occurs in a sinuous band stretching around the planet between 15° north and 45° south. It is possible by hypothesizing a suitable pole of rotation shift at some point in martian history to make this distribution coincide with an ancient 15° south latitude band. An alternative explanation of furrow distribution notes that it coincides with the edge of the south polar cap debris mantle, and with both the subsolar latitude at perihelion (the time of greatest solar heating) and generally low albedos. On this model, furrows are present throughout the equatorial region but are only exposed in their observed distribution through the concentration of wind erosion there.

The ages of small features like channels are very difficult to estimate. Most of the larger examples have some superimposed craters and appear to be at least 100 million years old. The small numbers involved make individual ages derived from crater counts very uncertain. It is perhaps safer to observe that most channels seem to predate the bulk of northern plains materials, which themselves appear to be about 1,000 million or more years old. The exceptional channels of the Tharsis and Elysium volcanic provinces may be the product of remobilized ground ice. Possibly most channels date back to a single great erosional event around 3.5 billion years ago.

Our best guess for the fluvial history of Mars would presently be as follows: A dense atmosphere formed early during or soon after planetary accretion. The atmosphere was initially warm and included a large amount of water vapour, but planetary cooling triggered extensive rainfall, producing the dendritic channels and crater furrows preserved in the martian southern hemisphere. Most of this rainfall percolated down through the impact-shattered outer kilometres of Mars and formed vast ground ice bodies as the temperature declined still further. Local melting, perhaps triggered by volcanism and by near-surface intrusion, produced the collapsed terrains and great outflow channels of the martian equatorial regions.

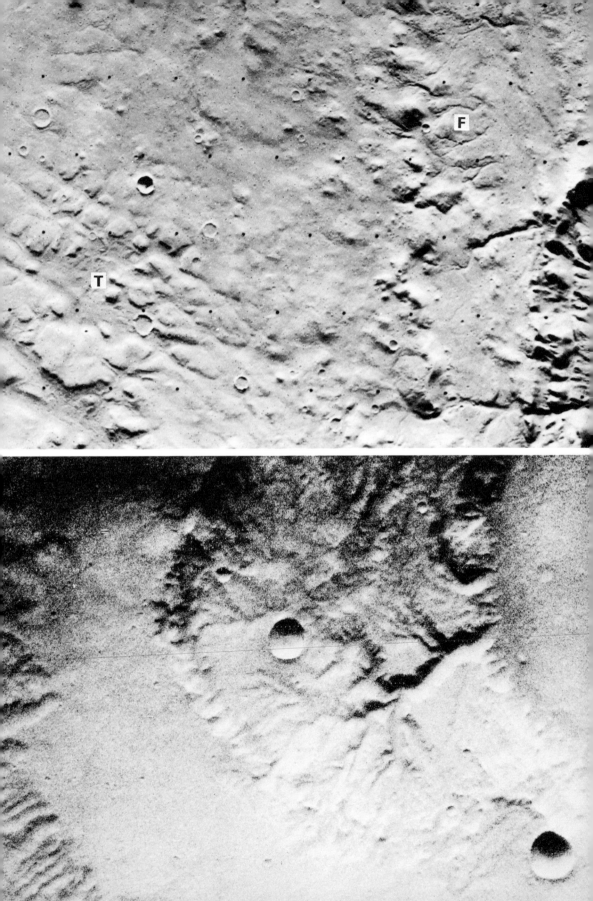

Mars: Fretted Terrain

Quite distinct from all the channel types we have discussed so far are the valleys of fretted terrain. Fretted terrain is one of several terrain types apparently formed by the disruption and partial destruction of heavily cratered rock units. It forms a transitional boundary between the cratered terrain of the martian southern hemisphere and the lowland plains to the north, in a 500km-wide zone around half the planet. As, among others, Bob Sharp and Mike Malin of the California Institute of Technology have described, it consists largely of flat-topped buttes separated by rectilinear valleys. Some of the valleys can be traced deep into southern cratered terrain. The picture opposite illustrates a quite typical area of fretted terrain, 60km by 40km. To the north, buttes progessively reduce in size to be replaced by rounded hills and knobs. In more detail, going from south to north through a region of fretted terrain, one typically encounters the following sequence: First there are a few linear fissures. Further north a simple pattern of intersection develops, and the pattern is usually rectilinear implying strong fault or joint control. As the pattern develops into a maze of intersecting linear depressions, the cratered terrain surfaces become subdued. Still further north the cratered terrain becomes reduced to isolated remnants, which often appear to have been tilted towards the north. These remnants become smaller and more widely spaced; eventually they are reduced to small nubbins, and finally disappear altogether.

No convincing terrestrial equivalent for fretted terrain has been proposed. It does not seem to have been produced by ordinary subaerial water erosion. One possible explanation is that fretting is largely the work of ground ice or groundwater sapping. In this, water seepage and plucking concentrated at the base of a cliff causes repeated collapse and cliff retreat. The uniform height of many scarps in fretted terrain supports this mechanism — the height may record an old water or ice-table level. Ground ice sapping would seem to be incapable of producing scarps, although a powerful mechanism for causing them to retreat. This may be the reason why fretted terrain is localized along the 'geologic equator', if, as we have proposed, great scarps were produced here by downdropping of cratered terrain in the north.

In some areas of cratered terrain, craters have provided multiple sites for the initiation of scarp retreat. In others, crater walls have acted as barriers to plantation, while their floors include the same jumbled blocks as chaotic terrain.

Mars: Albedo Features

Light and dark patches were seen on Mars from the very first telescopic observations, as noted earlier. As they were always of roughly the same appearance from year to year, they were soon identified as surface markings, rather than atmospheric effects. Having little else to work on, observers of Mars spent a great deal of time mapping these features and their changes, and trying to explain them. Patches changed shape and brightness, both with the martian season and more irregularly. Obviously the description of dark areas as oceans and bright areas as land was not tenable. Opinion remained divided between organic and inorganic accounts of the Martian surface almost until the first spacecraft missions. On the organic model, dark areas were regions of primitive vegetation, perhaps lichens, while light areas were barren deserts. Inorganic models were more varied, but many were based on the idea of the martian surface being covered by thin mantles blown around by the wind. One proposal was that volcanic ash was being continually erupted from the edges of some of the dark patches.

Mariners 4, 6 and 7 photographed a surface on which extensive vegetation was obviously out of the question. They failed to provide a complete answer to the problem of the albedo markings, both because of their limited coverage of the surface and because their photographic resolution was inadequate. Mariner 9 showed immediately that the atmosphere and surface dust can interact to a considerable degree. Its arrival at the planet coincided with a global storm which had raised dust up to 30km into the atmosphere and completely obscured surface features.

As the dust settled, the origin of the classical markings of Mars became clear. Syrtis Major, for instance, one of the most prominent dark markings, was resolved not into a single dark patch but into a very large number of small streaks, oriented in similar directions, similar to the 600km by 500km area shown opposite. Most other albedo features are also composed of streaks, some dark and some light, forming tails behind topographic features, commonly craters. It is clear that these features do not reflect deep-seated geological or topographic structure but merely demonstrate that the martian surface is covered in dust and sand stirred by the wind.

Three main types of albedo feature have been seen: bright streaks, dark streaks, and dark patches (called splotches) on some large crater floors. Some splotches have been resolved into dune deposits. It is natural, therefore, to suggest that dark areas have been areas of deposition, while bright streaks might indicate the stripping away of loose materials; or light areas might indicate deposition of materials in a different size range, or of different composition.

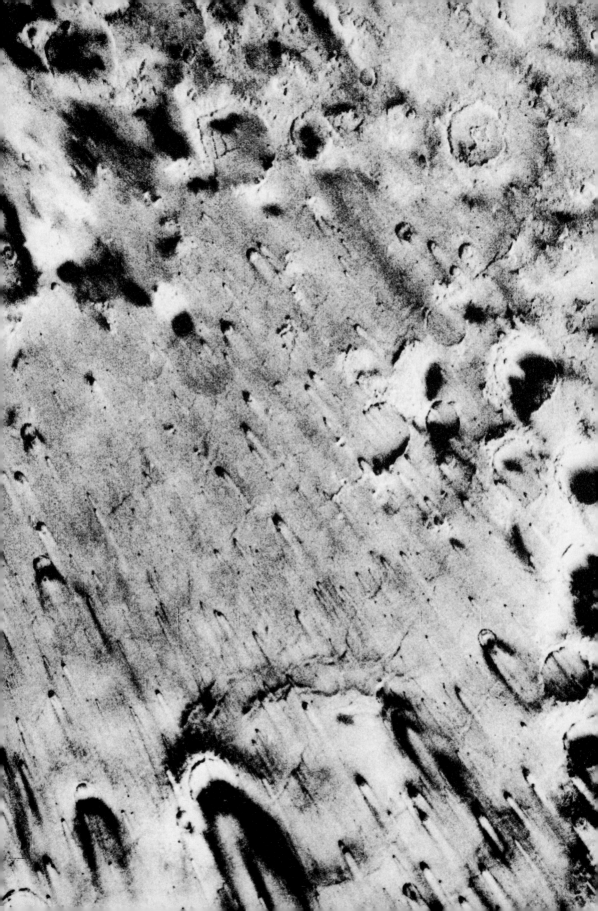

Mars: Variable Features

A great deal of effort during the Mariner 9 mission, particularly by Veverka and Sagan of Cornell University, was devoted to monitoring streak patterns for changes with time. The pictures opposite illustrate some of the variety among martian variable features, including bright streaks, dark streaks, compound streaks and a dark floor patch. Bright streaks were found to be stable for several months; dark streaks, on the other hand, could change considerably over a few days. Bright streak patterns did change between the flybys of Mariners 6 and 7 and the orbital observations of Mariner 9 and, again, between Mariner 9 mission and Viking. The greater stability of bright streaks seems best explained by them being composed of generally smaller particles than dark streaks. Somewhat surprisingly, wind is better able to move particles in the size range between dust and sand than larger or smaller particles than this. Deposition occurs preferentially behind craters and hills because these obstacles disturb the air flow over them, producing turbulent eddies behind them where low wind speeds allow dust to settle out. Streaks are most conspicuous behind small bowl-shaped craters. Larger craters with some portion of flat floor seem, instead, to have trapped material inside, and these craters are more commonly splotched than streaked. Streaks are most common on the relatively smooth level surfaces of moderately cratered martian plains. These plains are probably largely volcanic, and dark lavas under a thin loose dust cover would be ideal to display wind-produced disturbances.

Many attempts have been made, notably by Ron Greeley of the NASA Ames Research Center, to reproduce the characteristics of martian splotches and streaks in wind-tunnel experiments on Earth. With increasing sophistication more and more types of splotch and streak patterns have been successfully mimicked. Indeed, these wind-tunnel experiments have now been used to estimate wind speeds on Mars, together with the sizes of the particles involved and erosion and deposition rates. The experiments reinforce the idea that most light streaks are depositional, while many dark streaks are erosional.

Mapping of streak orientation suggests that dark and bright streaks are formed by different wind systems. Bright streaks appear to be formed by a return flow of dust from north to south after duststorms have spread dust all over the planet. Dark streaks may directly record the passage of local storms. Streaks near the poles are consistent with outward-spiralling winds driving a debris blanket before them. Several unusual dark streaks originate not at crater rims but from points on their floors. It has been suggested that they may be volcanic ash erupted from crater floor vents.

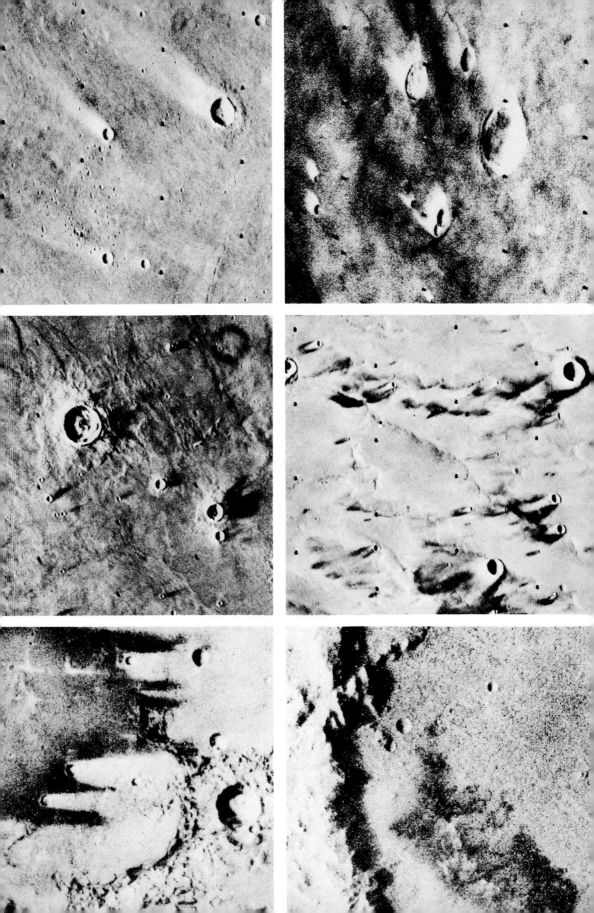

Mars: Wind Erosion

Few variable features (streaks and splotches) were observed to change during the primary Viking mission. This came as no surprise, as the season was expected to be one of generally low winds. Two exceptions, where wind activity was considerable, were the slopes of the Tharsis volcanoes and the interior of Hellas basin, both areas of steep slopes and, hence, unusually vigorous wind régimes. One important new observation made possible by the clearer atmosphere and Viking's better cameras was the detection of bright streaks with dark edges. These seem to have resulted from an episode of deposition, following one of erosion by winds from the same direction. In summary, Viking has put beyond much doubt the Mariner conclusion that albedo changes on Mars can be attributed solely to aeolian action and the deposition and sublimation of volatiles at the poles.

We have seen that wind is an important agent of sediment transport. Is it so effective as an erosional agent? This question is hard to answer, even for the Earth. The appearance of many terrestrial deserts, contrary to expectations, is the result more of very occasional water erosion and biological activity than of wind action. Only the very driest and climatically most stable deserts display many large features produced by wind erosion. Perhaps the most important features are streamlined hills or yardangs, investigated by Jack McCauley of the US Geological Survey, though even these are conspicuous only where wind has acted

for a very long time on soft rocks. Where sediments have been protected by a very thin layer of harder rock, wind activity is capable, once the top layer is penetrated, of eroding large pits and deflation hollows. The main picture opposite, 50km from top to bottom, includes several of the large irregular depressions, common near the first Viking landing site, from which material seems to have been removed by wind action. Aeolian action is more important on a small scale, where a whole variety of grooves and ridges can be eroded by strong recurrent winds. Examples of grooves and pits are provided in the lower left and middle right inserts, which are respectively 120km and 30km across.

We have already seen, in talking of crater streaks, that martian winds are capable of some erosion, and it is possible that the great dryness of the atmosphere over immense periods, when compared to the lifetimes of terrestrial deserts, has allowed wind action a dominant rôle on the martian surface denied to it on the Earth. The evidence so far is against this, however. The preservation of craters and other landforms dating back almost to the formation of the martian crust suggests that no martian erosional agency has been effective over large areas. Yardangs and pits and grooves *are* present on Mars, but they are not a dominant or even common feature. As on the Earth, martian wind erosion seems to have been effective only when operating on poorly consolidated soft rocks.

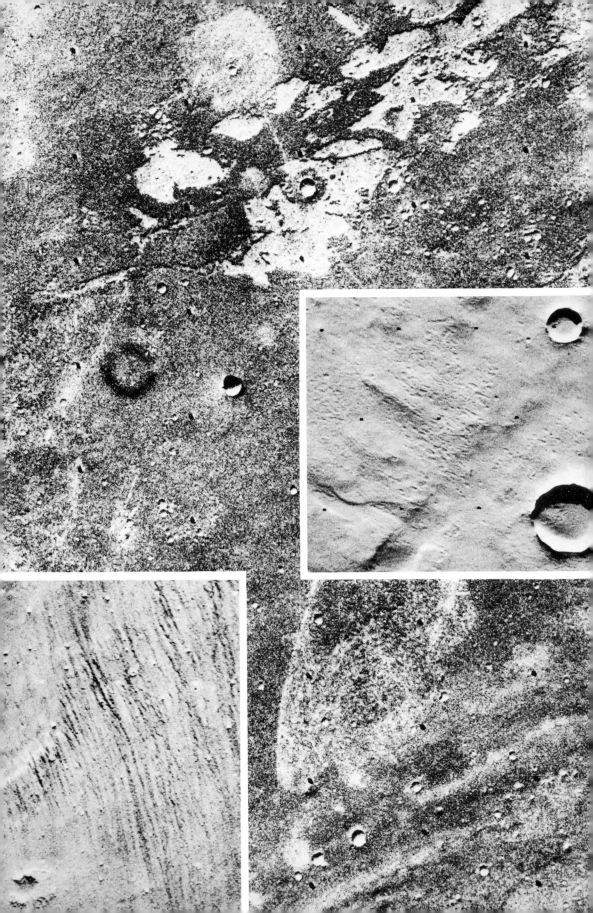

Mars: Other Wind Activity

As we have seen, Mars has a great variety of crater forms. One form, just recognizable from Mariner 9 images and documented thoroughly from Viking, is the pedestal crater. It is probable that some, and perhaps many, pedestal craters result from wind removal of material from between craters. Underneath craters, similar material may remain undisturbed, protected by a covering of impact ejecta.

At least two kinds of aeolian deposit are important (this ignores for a moment the many episodes of deposition and subsequent transport suffered by thin, mobile mantles). The layered deposits of polar regions are discussed on pages 176 and 178. Sand dunes have been resolved, both in a large northern circumpolar belt and within a number of 'splotches' on crater floors. The picture opposite shows a typical 'splotch' within an 80km crater, part of which is seen in the bottom right corner. The 'splotch' consists of a field of dark sand dunes, arranged in transverse ridges very similar to dune configurations in many deserts on Earth.

Unlike most of the other agents responsible for the present surface appearance of Mars, wind activity has been observed at close range by the Viking landers. The lander cameras have provided several lines of evidence for fine-particle motion. Extensive drift accumulations are present at the first Viking landing site. A number of rocks at both sites have bright deposits in linear streaks behind them. The directions of these streaks are almost identical to those of the much larger features visible from orbit. Although most of the rocks themselves look remarkably fresh, a few examples of more rounded and possibly sand-abraded rocks and pebbles are visible. Some small motions of materials have been observed between pictures of the same area taken several days apart. These motions corresponded to the greatest wind speeds measured by the landers.

In general, the Viking landers observed very repetitive daily patterns of wind, temperature and pressure. Winds were light initially (during the local summer) but became more irregular with the onset of autumn. Atmospheric pressure declined quite steadily right through late summer and autumn in response to sublimation of carbon dioxide onto the north seasonal polar cap. During the course of the first year of operation, several local duststorms were observed from orbit to pass close by the landers. However, all that were observed from the surface were darkenings of the sky.

Mars: The Polar Caps

The white polar caps of Mars are almost the only features that can be monitored reliably from Earth. They were seen, and correctly ascribed to ice deposition, almost as soon as telescopes were turned towards the planet, and it was soon noticed that they grew and diminished with the martian seasons. The southern cap has received more attention than the northern one through the accident that it is largest when Mars passes closest to the Earth. The picture opposite shows part of the permanent south polar cap (C), and the extensive winter frost (F). Hellas basin is visible at H.

Since their discovery, argument has continued about the composition of the cap deposits. By analogy with the Earth's caps they were initially assumed to be water ice. An alternative carbon-dioxide ice composition gradually found support, however, and the abundance measurements of carbon dioxide in the atmosphere made it plausible. Physical conditions of the martian atmosphere measured by spacecraft proved the extended caps of polar winter to be largely carbon dioxide frosts.

Mariner 9 made extensive observations of both caps during its year in orbit. Like those from the Earth, however, they favoured the southern cap. The northern cap was shrouded in haze early in the mission and was not approached so closely by the spacecraft. The extended northern cap was seen to be quite symmetric about the pole and grew and retreated in a regular way. The southern cap, in contrast, although in winter circular and centred on the pole, retreated very irregularly and by martian summer was centred several degrees away from the pole of rotation. The difference in behaviour is mostly the result of much more irregular topography beneath the southern cap.

Both caps stop retreating in late summer. This stability has been used to argue that the residual caps are of water ice, rather than of carbon dioxide. It is also possible to explain the stability using only the action of wind and dust on carbon dioxide ice, and Viking seems to require yet another about turn — considerable differences between the residual southern caps observed by Mariner 9 and Viking, together with temperature measurements, imply that although the inner northern cap is water ice the residual southern cap is carbon dioxide ice!

Beneath both frost caps, Mariner 9 observed striking and unexpected features. Foremost among these were curvilinear bands, dark because free of frost. High-resolution pictures revealed that many of these bands were lined with parallel systems of stripes. They looked very like glacial moraines. Water-ice glaciers are not at present possible on Mars, however: polar temperatures are much too low for ice to flow. As more and more pictures were taken of the southern polar region it became clear that the apparent striping was the result of seeing the edges of a large number of very thin layers. Since that time, most people have argued that these layered deposits are evidence for cyclical sedimentation and constitute the only nonterrestrial evidence to date for repetitive climatic change over long periods.

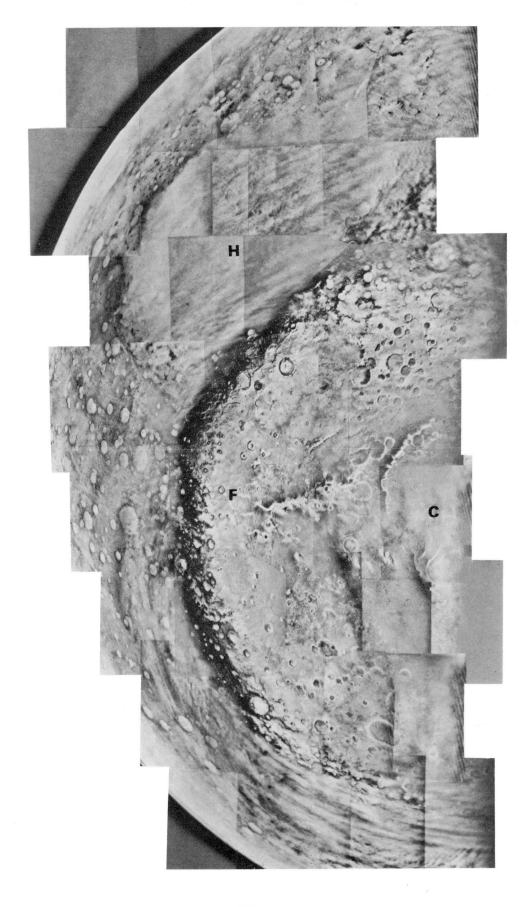

Mars: Polar Stratigraphy

The stratigraphy of the martian poles is fairly clear. At the south pole, for instance, the polar cap lies above layered deposits which, in turn, lie over massive unlayered etched plains materials and a heavily cratered bedrock. In the picture opposite, P is the polar cap, L are layered deposits, and E are etched plains. The nature of the two groups of thick deposits, the layered and unlayered, is unclear. Their confinement to polar regions indicates meteorological control of their deposition, a strong argument that they are wind-transported debris. Their materials are probably derived from equatorial regions — perhaps the canyonlands or fretted terrain.

Two rival explanations were immediately proposed for layered deposits. One made them depositional plates whose locations were determined by the position of a wandering pole of rotation. The geometry of layered deposits is better accounted for by recent erosion of more extensive layered successions. A number of other models for layering have been developed since Mariner 9, including one invoking carbon dioxide glaciers.

If it is accepted that cyclical deposition has been responsible for layered deposits, these cycles must record climatic changes, forcing episodic carbon dioxide sublimation and/or water ice precipitation. The most likely cause of climatic variation would be the periodic orbital changes to which Mars is subject, and particularly changes in the spatial orientation of its poles (obliquity). The Earth has quite recently been shown to be very sensitive to such changes and Mars is subject to a much greater range of changes. It is suggested that at high obliquity the poles would be warmer than at present and ices would be less stable. At low obliquity, both carbon dioxide and water ices would be laid down, the latter producing the layered terrain, with carbon dioxide subliming from the atmosphere and water leaving the regolith. At the present intermediate obliquity, water would move from one pole to the other with the precessional cycle. The terraced edges of layered terrain may be maintained during deposition either by wind erosion or by preferential deposition on level surfaces and inhibited deposition on equator-facing slopes.

No model of layered terrain development is yet close to general acceptance. A serious problem to be faced by the one we have just outlined is the Viking finding that the southern residual cap is of carbon dioxide and not water ice, which implies that the production of the layers may have little to do with water ice deposition. Another problem is that Viking has observed at least two sets of terraces requiring two periods of cyclical deposition, separated by a period of erosion. Such large-scale cycles are difficult to understand.

Beneath layered deposits at the south pole lies a thick, comparatively old, deposit which appears to have suffered considerable wind erosion, being covered in pits and etched regions. The northern pole has a similar unit, but it lacks such prominent pitting. Both units are presumed to lie over ancient heavily cratered terrain.

Mars: Polar Landforms

One of the most important observations made of polar deposits is their extreme youth, compared to the rest of the planet's surface. Over an area of almost 1 million square kilometres no fresh craters are visible. Whether craters are being buried or eroded, high obliteration rates must be operating. Layered deposits may be only of the order of 100 million years old.

Recent Viking pictures have revealed that the curvilinear bands in layered terrain discussed earlier (page 176) may not be scarps at all, but troughs! This implies that the large-scale topography of the layered terrain is undulating rather than stepped. In between troughs, the terrain can be described as 'egg-carton'. The idea of differential accumulation on level areas and slopes has been restored to prominence to explain the maintenance of these troughs. Wind scour may also help to keep troughs clear, as wind speed should be highest in the troughs causing most erosion, deepening the troughs, and tending to concentrate winds still further. The 'egg-carton' pattern seems to result from the intersection of different trough systems to produce an interference pattern. The different systems may be centred on past pole positions.

Another Viking surprise is that around the north polar cap lies the most spectacular dune field of the Solar System, which has been intensively studied, notably by Jim Cutts. It covers more than 5 million square kilometres, an area greater than the Sahara and Arabian deserts combined. Buried and revealed annually by the advance and retreat of the seasonal polar cap, its interior is dominated by vast arrays of transverse dunes, with barchan dunes appearing at its edges. Part of this dune field, which measures 100km from top to bottom, is shown in the picture opposite. Transverse and barchan dunes imply that little movement is occurring. If the flow of material was high, longitudinal dunes would be more prominent. The winds inferred from the dune geometries are the ones predicted for the polar regions — Coriolis winds spiralling out from the pole.

The great problem with the circumpolar dunes is to explain how so much sand can have reached the area. Layered deposits appear to have been emplaced from suspension in the atmosphere and must be composed of fine dust and ice. Dune-forming sand, on the other hand, is quite coarse and cannot, therefore, have come directly from layered deposits. It is possible the dunes are composed of dust aggregates. They may, alternatively, be derived directly from more equatorial latitudes.

Layered deposits include a number of very bright streaks informally called 'searchlight' features. These may be longitudinal snow dunes.

Finally, the south polar region of Mars contains a number of unexpected features, which have been studied by Jim Cutts of the Planetary Science Institute, Larry Soderblom of the US Geological Survey, and others. The rocks and underlying layered and etched deposits appear to include a number of lava plains and many large cones are visible. Narrow ridges are also common and have often been explained as dykes. An alternative, more radical, explanation is that some of them are eskers — a record of stream flow under great thicknesses of ice.

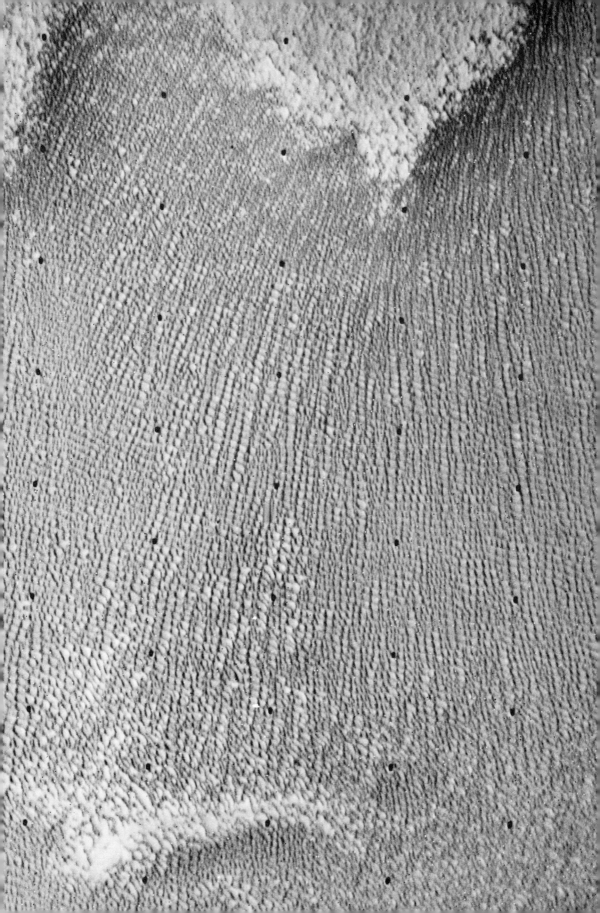

Mars: The First Viking Landing Site

The initial choice of sites on which to land the Viking spacecraft was made on Mariner 9 data before Viking was even launched. The prime aim of Viking was to look for life on Mars, using a series of highly sophisticated experiments housed on the lander spacecraft. Sites were chosen in the northern hemisphere for a number of reasons. Firstly, Mariner 9 had suggested that these regions were smoother than most areas in the southern hemisphere. They were also topographically lower, and thus the atmospheric pressure was higher near the ground, and this higher atmospheric pressure would assist in decelerating the spacecraft by parachutes. The Chryse Planitia, a vast extent of plains in a basin-like embayment on the edge of the northern cratered terrain, was considered to be one of the best sites; not only did it appear to be smooth but running into it were numerous large channels interpreted as the beds of ancient rivers that brought sediments into it. The possibility of water in this region, at least in the past, was considered to enhance the chance of there being life.

The map opposite shows the Chryse basin and, as can be seen from the contours which are marked in kilometre intervals, the floor of the basin is at minus 3km. Before Viking arrived at the planet it had been tentatively decided to land the spacecraft in the southern part of the basin. However, once pictures of this region started coming back from the orbiter it was clear that in this region not only were there many impact craters but also that the terrain was more rugged than expected, showing signs of fluvial channelling. Blankets of impact craters are usually blocky, and large blocks could be a hazard to the landing spacecraft.

It was decided to change the orbit of the spacecraft towards the northwest to look for a suitable site in the middle of the Chryse Planitia. The orbital pictures showed a terrain that appeared relatively smooth but, unfortunately, radar data acquired from there suggested that the surface in the centre of the basin could well be rough or blocky; and the search moved west to the region of the mouth of Kasei Vallis.

A photomosaic of this region is shown opposite below. On the western side of the mosaic (left) we see striations associated with the mouth of Kasei. However, towards the east there are relatively smooth areas cut by low ridges, giving the whole area the appearance of the lunar *maria*. *Mare* ridges on the Moon normally have slopes of only a few degrees, and, on this basis as well as in the knowledge that the lunar *maria* would be a safe place to put the spacecraft down, it was decided to land Viking 1 at the spot marked with a star on the photomosaic. (The area of the photomosaic is shown boxed on the map.)

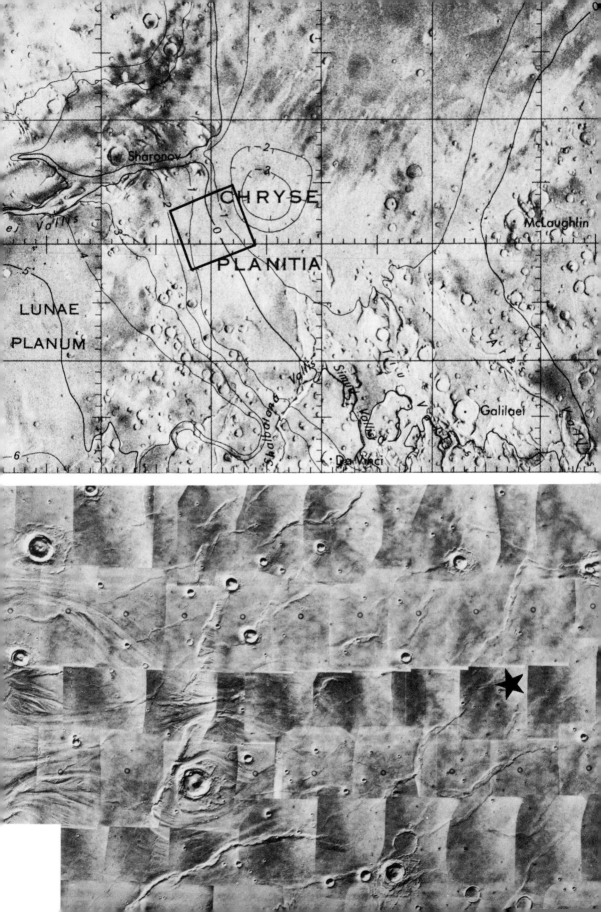

Mars: The Viking 1 Landing Site Area

The first view to be had of the martian surface from the landing spacecraft was an exciting experience, especially as the pictures were so sharp. The first picture to be returned was a close-up view of the material immediately surrounding the spacecraft footpad; the picture was taken so soon — only 25 seconds — after landing that the cloud of dust thrown up by the spacecraft as it came down is still visible, causing the slightly fuzzy appearance on the left-hand side. The picture is made up of a series of vertical scans from left to right, so that, by the time the bulk of the picture had been taken, the dust had settled. This dust can be seen resting in a hollow of the footpad on the right-hand side. The surface is seen to be rather more blocky than had been expected; the largest block in the middle of the picture is about 10cm across.

Panoramas around the landing site looking towards the horizon show that the whole area is blocky, and some blocks such as the one (nicknamed 'Big Joe') on the left-hand side of the panorama are as much as 2m across. The scene is very much like a typical rocky desert, consisting of many blocks surrounded by fine-grained material; in the case of Mars, the fine material is of dust rather than sand-grain size. The dune-like features visible in the middle distance are more like snowdrifts than sand dunes, apparently having been scoured by winds. The nature of the irregular skyline is uncertain, but we may be looking at the rims of small impact craters that occur on the surface, as shown in the picture on the previous page.

Many different types of rock are observed on the surface. In some of them crystals can be seen and many of them have vesicles, suggesting that they are of volcanic origin.

How do we interpret the geology in this region? At some stage in the history of the region material has been carried down the extensive valleys to deposit sediment in the basin. On the other hand, by comparison with the Moon, the terrain looks remarkably like the volcanic lunar *maria*. Are these two views consistent? One interpretation is that the Chryse Plains are underlain by alternating deposits of sediment brought down from the surrounding high ground around the basin by occasional catastrophic periods of flooding and volcanic lavas produced by equally catastrophic flooding by fluid lavas across the floor of the basin. It appears likely that in the region of the landing site the last episode was a volcanic one. This would suggest that the blocks on the surface are representative of an underlying lava flow, rather than being material carried down from the higher regions surrounding the Chryse Plains by floods.

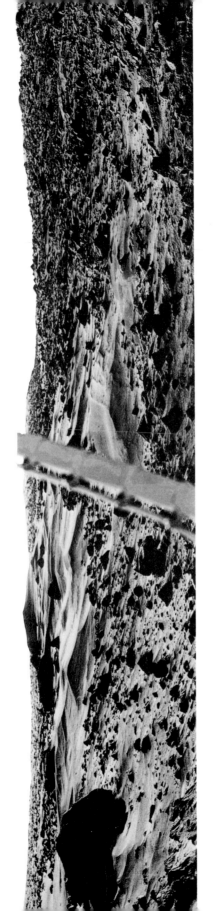

Mars: The Search for a Second Landing Site

Having got one lander down safely, there remained the difficult task of finding another good landing site. Extensive searches in a number of regions in the northern hemisphere proved abortive, the terrain being pockmarked with craters and in places rugged. One of the last areas to be searched was the Utopia region (towards the top right of the map on page 107).

In the top picture opposite we see a region south of Utopia known as Nilosyrtis Mensae. Here we see clearly the cratered highlands where they have been broken down by erosive processes. The highland terrain is cut by numerous valleys that tend to follow regional patterns presumed to be associated with a tectonic pattern, and the floors of these valleys are filled with material characterized by striations. This material is considered to be debris that has been eroded from the valley sides and has been transported down the valleys as mass-wasting products. As erosion continues, the high ground between the valleys is progressively removed, leaving a jumbled plain.

As we move north from the southern hemisphere terrain into the plains we encounter numerous landforms that are, to say the least, bizarre. Many of them are extremely difficult to interpret in terms of our knowledge of geological processes on Earth and they may be the result of permafrost activity, wind erosion and deposition, and volcanism. One example of a possible permafrost feature is shown in the middle picture opposite, where we see what might be called a 'thumbprint texture', one explanation for which is frost heaving in the near-surface layers. These features, and many others, are, however, still enigmatic and require more detailed study.

One of the few relatively smooth areas in Utopia is the one shown in the bottom picture and it was this region that was chosen for the landing site (marked by an X). In this region, polygonally cracked terrain with craters appears to have been mantled by a thin overlying layer, giving the terrain a subdued form. Whether this is wind-blown material or related to the nearby 100km-diameter crater Mie is not known, but the mantled terrain augured well for a safe landing. You will notice also that in some areas there is a curious pitted terrain that was considered by some to consist of small sand dunes. It was in this pitted terrain that Viking 2 landed and, as we shall see on the next page, there is no evidence either of sand dunes or any other phenomena that could cause this peculiar terrain. None of the craters look fresh; most have wide pedestals, and may be degraded rampart craters.

Based on geological interpretations of photographs taken from orbit, the geology in the second landing-site area is quite different from that in the Chryse basin, and geologists such as Tim Mutch, leader of the Lander Imaging Team, and his colleagues awaited with eager anticipation the first views of the landscape in this new area.

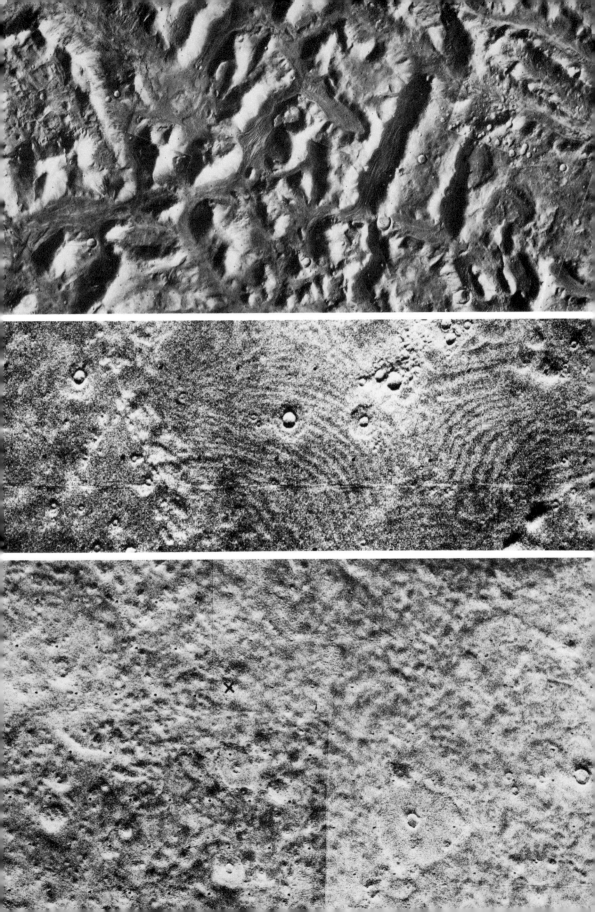

Mars: A Landing in Utopia

The first pictures from the second landers were a surprise because they looked so similar to those from Lander 1, despite the different geological environment. The biggest surprise of the Lander 2 pictures was that, if anything, the landing site was more blocky than the area around the Chryse site. In fact, the spacecraft landed on one of the blocks so that the craft was tilted over by 8°. There is, however, one fundamental difference between these two sites, and that is that in the Utopia region the skyline is level. Thus the interpretation based on Orbiter pictures that the area is relatively smooth was confirmed; but the fact that so many boulders are present had not been predicted.

Examination of the pictures on this page show clearly that, here again, many of the blocks are vesicular and may be interpreted as volcanic. Dust drifts are not prominent, but there is plenty of fine-grained material between the blocks on the surface. The picture on the left (turn the book sideways) shows the sample arm being used to collect some indurated surface dust lying between some rock blocks. It was not sure whether the sample arm would collect this 'crusty' material but, as we can see in the adjacent picture, where the arm has been removed, there is a shallow groove indicating that sample material had been moved by the collecting arm. Analysis of all the samples collected, in both Utopia and Chryse, show that surface material at these sites is strongly weathered.

The picture at the top right shows the surface near the spacecraft. Of particular interest is a linear trench which is about 10 to 15cm deep and can be traced for more than 10m. The trench is partly filled with sediment, which is finer than that on the adjacent surfaces. The trench forms part of a more extensive polygonal network seen around the lander. The origin is not certain, but it may be the result of cyclical freezing and thawing of groundwater. Similar structures form in periglacial regions on Earth where they are a result of water freezing in the soil to form ice wedges. If this interpretation is correct, these features must have formed at a time when liquid water was stable on Mars and the atmosphere was quite different from what it is today. Features like this were not seen in Chryse, perhaps because Chryse is at a somewhat lower latitude where conditions may have been more temperate.

It is difficult to explain why, despite the totally different geological environment of the two landing sites, they should, with a few exceptions, be so similar. This is particularly important with regard to interpretations about the origin of the boulders scattered over the surface. One explanation for this is that, at each of the sites, there are impact craters and it is possible that most, or all, of the blocks surrounding the two spacecraft owe their origin to having been excavated by impact cratering events. On the Moon, of course, blocks on the surface would be broken down by continued bombardment by micrometeoroidal particles hitting the surface, but on Mars — because it has an atmosphere — the surface is protected from attack by the small particles which, as they enter the atmosphere, burn up.

Mars: The Moons of Mars

Once Galileo had shown that the Moon, Earth's satellite, was not a unique body by identifying four satellites revolving around Jupiter (there are now known to be at least twelve), there was every reason to suppose that other planets might also have moons. Kepler surmised that, since Venus had no moon, Earth one and Jupiter four, fitting these to a geometric progression implied two moons for Mars and three for a hypothetical planet between Mars and Jupiter — oddly enough coinciding with the asteroid belt. And in 1726, in *Gulliver's Travels*, Jonathan Swift elaborated on the two hypothetical martian moons, describing the inner one as travelling quickly through the martian sky and the outer one more slowly. These conjectures proved substantially correct, though for all the wrong reasons.

The real search for the martian satellites was begun by Herschel in 1783 and continued by other astronomers until, in 1877, when the planet was in a particularly close position for viewing from Earth, a US astronomer, Asaph Hall, found first the outer moon and then a week later the inner moon. These he named respectively Deimos (Terror) and Phobos (Fear), after the two sons or attendants of the god Mars in Homer's *Iliad* . . . although perhaps Hall should have named one after his wife who, reputedly, revived his flagging spirits, no doubt with a cup of tea, after a long and initially fruitless search.

Only one hundred years after Asaph Hall's discovery three spacecraft (Mariner 9 and Vikings 1 and 2) had approached the satellites close enough to take pictures and record other important data. Before this time little could be discovered about the moons from Earth because of their small sizes — even these had to be worked out from observed brightnesses and assumed albedos.

Before Viking, Mariner 9 had determined that both satellites are irregularly ellipsoidal in shape, roughly 1.5 times as long as they are wide, Phobos (top and insert) being about 27km in its longest dimension and Deimos (bottom) about 12km. It was confirmed that the rotation of each moon is locked into synchronization with its revolution around Mars (one rotation during one revolution); thus one side of each body always faces the planet, with the longest axis pointing towards Mars, just as the longest axis of the Moon points towards Earth, and for the same reasons (see page 58).

Both Phobos and Deimos were seen to be heavily cratered and to be covered in a dark grey regolith, although Deimos appeared smoother than Phobos. The largest crater on either moon is Stickney (10km diameter), on Phobos. The impact which formed this crater knocked a substantial chunk off the moon, as can be seen in the insert at upper right.

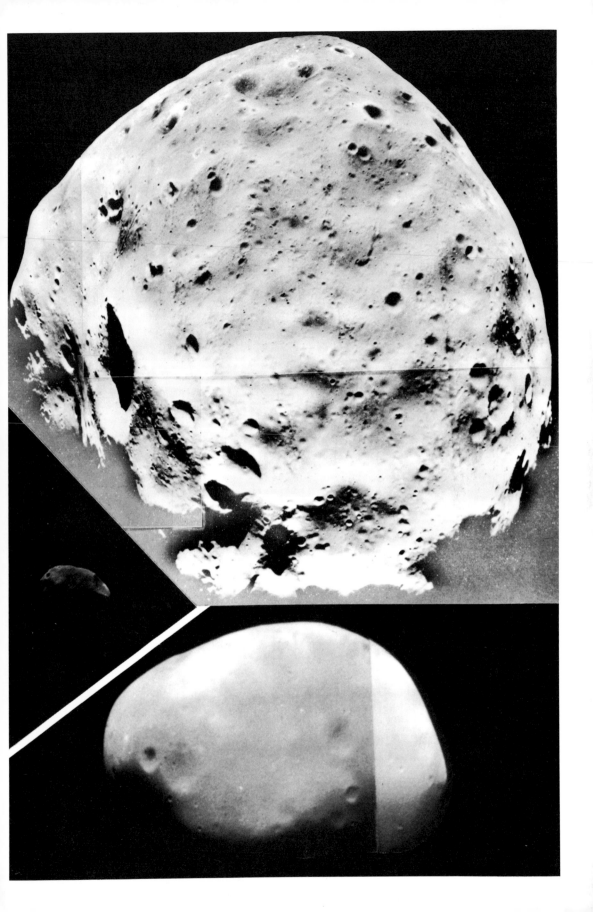

Mars: The Moons' Motion, Cratering and Albedo

Mariner 9 determined that the inner moon, Phobos, orbits at a distance above the surface of Mars of only about 6,050km, which makes it readily visible from the surface: it has actually been photographed by the Viking 1 lander. Due to its quick orbital period (7h 39min) it can be seen crossing the pink martian sky two or three times daily. Because it revolves in the same direction that Mars rotates, it rises in the west and sets in the east (moving the opposite way to the apparent path of the Sun in the martian sky), taking 11h 6min before it passes over the same spot, since it has to catch up on the amount the planet has rotated during the moon's orbit. Just how close Phobos is to the surface can be seen in the strip mosaic on the right where the black blob of the moon can be seen as it moves between the Viking orbiter and Mars. The mosaic at upper left shows the shadow of Phobos cast on the planet's surface.

Deimos, on the other hand, orbits about 21,000km above the surface and this, together with its small size, would make it look like a bright (but moving) star from the surface of its mother planet. At 30h 18min its orbit takes a bit longer than a martian day (24h 37min), so that it is about 5.5 martian days before it is seen in the same part of the sky again. Both moons revolve close to the plane of the planet's equator.

Mariner 9 confirmed the intriguing possibility that the orbit of Phobos is decaying, that the satellite is slowing down and getting closer and closer to the surface of Mars. It is reckoned that, if this process continues at its present apparent rate, Phobos will fall from the martian sky within the next 100 million years—a mere tomorrow on the geological timescale.

It was once suggested, based on Phobos' motion, that it was a hollow body—perhaps even an alien artificial satellite! The unlikelihood of this can be judged from the lower left picture, which shows a close-up of the cratered surface of Phobos; the picture has been specially processed so that albedo differences are enhanced. This process enables one to see, clearly, dark patches in the floors of the craters, especially the largest. Mission scientists have investigated the photometry of these patches and conclude that the dark material is different from the surrounds in texture rather than in composition. They suggest that this implies impact melt, especially as it is seen best in the younger craters, but alternative explanations could include subsurface layering, or gravity sorting of coarse debris in craters. Crater counts show that both Phobos and Deimos have ancient surfaces that may be over 3,000 million years old.

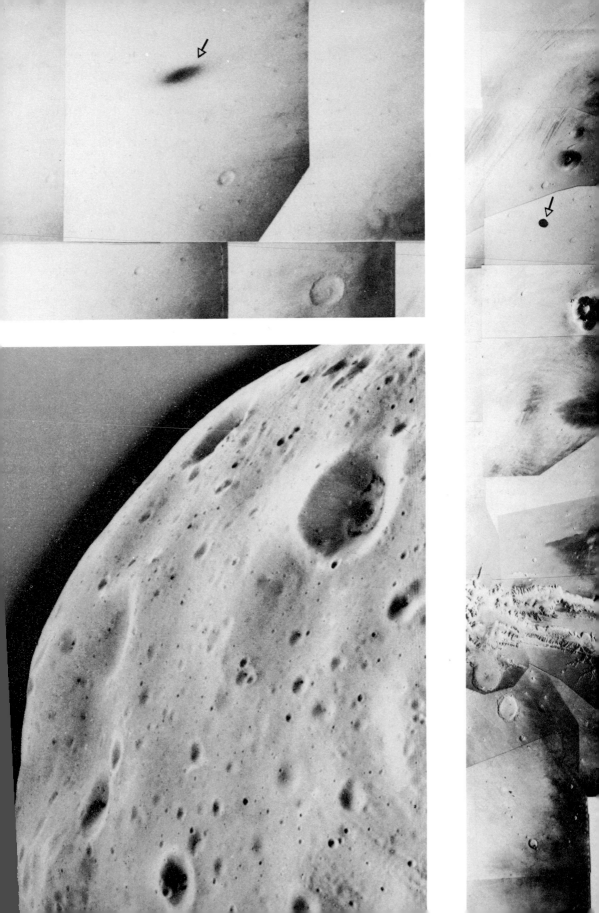

Mars: The Grooved Moon

The data gathered by Mariner 9 was confirmed and refined by the two Viking missions. For example, the orbits of the two moons were worked out more exactly by photographing them against background stars. Because the Viking orbiters had twin imaging systems the difference in brightness between stars and moons could be accommodated. Flybys were so close (23km from Deimos in one case) that mean densities could be worked out from the gravitational pull on the spacecraft. For Phobos this was combined with the results of an analysis of the moon's spectral reflectance seen from Viking 1 lander images to conclude that the moon is composed of a dark low-density material very similar to a type I or II carbonaceous chondrite meteorite. According to some theories, such bodies could only have formed in the outer half of the asteroid belt, so that Phobos may be a captured asteroid.

The most startling revelation of the Viking photographs is that Phobos is criss-crossed by long grooves, usually 100-200m wide and about 20m deep (above). Distribution maps have been made which seem to show they are related to the 10km crater Stickney, and it was suggested that they might be the surface expression of fractures produced in the impact. It is certain that the meteorite which formed Stickney must have been about the largest Phobos could accomodate without splitting apart.

Nevertheless, Stickney is at one extreme of Phobos, so the cracks could possibly, instead, be caused by tidal interaction with Mars. Phobos is certainly close to the Roche limit for its density: this means that, were Phobos a liquid body with no cohesive strength, it would be torn apart by tidal forces.

Under high resolution the grooves can be seen to be formed from chains of craters. They do not look like secondary impact crater chains and the craters have been explained as the result of loose material sifting into cracks, or steam escaping from the interior. Carbonaceous chondrites usually have some water content; water inside Phobos might have been heated by the Stickney impact and have escaped through fractures generated during the Stickney impact.

Deimos' surface (below) is different from Phobos' in two main respects. Firstly, no grooves are visible and, secondly, the surface is covered with a smooth, light material which partly fills most of the craters. These factors combine to give a smooth-appearing surface. Among all the fine grained debris can be seen house-sized blocks, probably thrown out by meteoritic impact. It is not clear why Deimos should differ so much from Phobos. It may be that different amounts of impact debris were produced because it is made of different materials, although the limited information available does not indicate much of a compositional difference. In any event, it is clear that these probably undifferentiated bodies are important in understanding the origins of the planets.

Venus: The Shrouded Planet

Venus is of particular interest to planetologists because of its broad similarities to the Earth—its mass, diameter and mean density, for example, are almost the same as those of Earth. It is also the closest planet to Earth, occupying an orbit between those of Earth and Mercury, and because of this and its cloudy atmosphere it is one of the brightest objects in the night sky.

But there are also many differences between the Earth and Venus: Venus has no moon and no magnetic field, and it rotates very slowly on its axis, in a retrograde sense. Thus the Sun rises in the venusian west and sets in the east, the whole venusian day lasting 243 Earth days. However, it is unlikely that anyone on the surface would be able to see the Sun rising, because of the cloud or haze layers which shroud the planet. The clouds are thought to be composed largely of sulphuric acid droplets, while the regions which show up dark in ultraviolet light (see Mariner 10 ultraviolet photomosaic opposite) are thought to be rich in sulphur. It is thought that the sulphur may be produced by photodissociation of carbon-oxygen-sulphur compounds from the lower atmosphere, this being only one reaction in a complex and poorly understood cloud chemistry.

The clouds are contained in a massive atmosphere of carbon dioxide which produces a pressure at the surface about 90 times that of the Earth's atmosphere at sea level. The temperature at the surface is an incredible 480°C.

It is considered that, because Venus formed closer to the Sun than the Earth, its atmosphere was sufficiently hot to prevent condensation of water, which would have absorbed some of the carbon dioxide and provided a cooling, stabilizing influence. Thus, since carbon dioxide lets heat from the Sun *into* the atmosphere but does not let it *out* so readily, a 'runaway' greenhouse effect was produced.

It is supposed that in the colder outer atmosphere most of the planet's original water has been photodissociated by solar ultraviolet radiation into oxygen and hydrogen. The hydrogen has been lost to space, and the most straightforward way of accounting for the excess oxygen is to suppose that it reacted with the surface rocks; however, it has been suggested that, without the action of running water to continually expose fresh rock for oxidation, sufficient oxygen could not have been taken up in this way.

Fresh rock at the surface might be provided by a process like that of plate tectonics on the Earth, but the evidence available so far is not adequate to show if Venus has Earth-like tectonics or not. If a planet apparently so similar to the Earth in many ways does not share the same tectonic régime, it raises the important question as to why the Earth's régime is the way it is.

Venus: Venera and Pioneer

The venusian clouds stop us looking at the surface directly, but radar images have been made of the planet by the radio-telescopes at Arecibo, Puerto Rico, and Goldstone, California. The upper image, from Goldstone Tracking Station, shows an area approximately 1,500km across. A large canyon (1,500km long, 150km wide, 2 to 4km deep) can be seen at the centre running from top to bottom (NNE-SSW) and other features, notably possible impact structures, have also been detected. The chasm, by analogy with the Rift Valley of Africa and the Valles Marineris on Mars, probably results from extensional tectonism. The range of elevations on the surface is about 5km compared with about 20km on the Earth and 30km on Mars.

Two pictures have been sent back from the surface of Venus by the Soviet Venera 9 and 10 spacecraft which landed on the planet in 1975, 2,000km apart. A part of one of these pictures is shown below; this demonstrates that there is daylight beneath the clouds, and in this light can be seen a panorama of angular to rounded blocks of the order of half a metre across, separated by finer material, stretching away from the Venera 9 lander. The lander instruments show that the blocks have a density of between 2.7 and 2.9 g/cm³, which is typical of terrestrial basalts, as were the radioactive content of the rocks beneath both landers, and the low albedo.

The presence of basalt would indicate that the planet has been hot and that it has differentiated into a crust, mantle and probably a core, although how volcanism is manifested and how lava behaves under such extreme conditions is open to speculation.

In December 1978, during the writing of this book, a Pioneer spacecraft released four probes from a Venus orbiter; these were followed after a few days by the Soviet Venera 12 spacecraft. The four Pioneer probes were designed to penetrate the planet's atmosphere at various points, gathering samples and taking readings until they hit the surface. In the event, one survived the impact to send back data for 110 minutes. Preliminary results support many earlier observations and suspicions and also indicate some surprises such as the presence of violent lightning spearing through the night sky; the atmosphere is warmer at the poles than at the equator; and over the north pole (and possibly the south pole) is a 600km diameter gap in the clouds. Perhaps the most important discovery was by the mass spectrometer and gas chromatograph, the presence of very large amounts of argon-36, an isotope which was thought to have been largely lost from the terrestrial planets very early in the history of the Solar System, being blown away by the Solar Wind. A large Ar_{36} content in accepted theory would mean that Venus evolved as a planet much later than did the other terrestrial bodies.

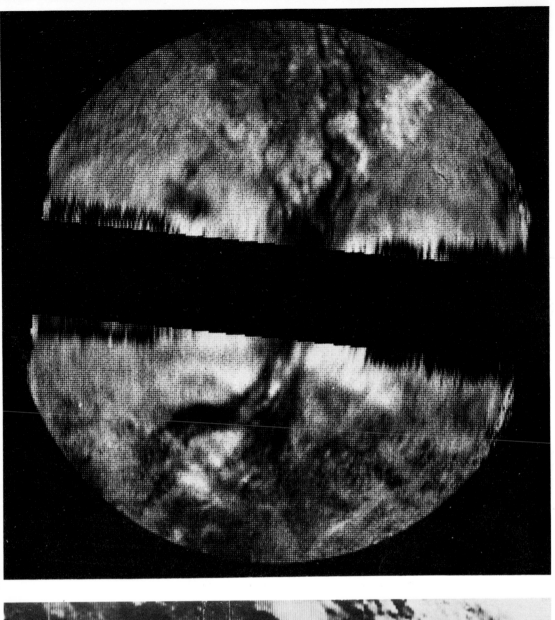

Satellites of Jupiter: A General View

So far in this book we have examined the Moon and the terrestrial planets—those rocky bodies closest to the Earth in size—which are quite different from the giant planets of the outer Solar System, with their thick gaseous atmospheres. Yet there remain a further seven bodies in the Solar System of comparable size: they are Io (3,652km), Europa (2,900km), Ganymede (5,270km) and Callisto (4,500km), all satellites of Jupiter; Titan (5,830km), a satellite of Saturn; Triton (about 4,000km), a satellite of Neptune; and the planet Pluto (about 6,000km). These sizes range from nearly the diameter of Mars down to rather less than the diameter of the Moon.

Four of these bodies are being explored as these words are being written, for the first close-up pictures of the four largest satellites of Jupiter—Io, Europa, Ganymede and Callisto—have just been taken by the US spacecraft Voyager 1.

These satellites are small and indistinct when seen through telescopes from the Earth. Under the best seeing conditions with large telescopes, light and dark markings can be seen on their surfaces; along the top are drawings by Earth-based observers of (from left to right), Io, Europa, Ganymede and Callisto, Io being the nearest satellite to Jupiter and Callisto the furthest. Already, from these drawings, the broad characteristics of each satellite can be seen: Io has dark poles and a bright equator; Europa is less 'contrasty', with shaded equatorial regions and lighter poles; Ganymede has light and dark spots all over its surface; and Callisto is much darker than the others, but with a few light patches. As seen from Earth, all four satellites are frequently eclipsed by Jupiter as they revolve about it, and of course they also pass in front of, or *transit*, the face of Jupiter. Satellites in transit provide a favourable opportunity for observing markings, and a transit of the satellite Ganymede is shown in the series of drawings down the right-hand side.

As the Voyager 1 craft neared Jupiter in March 1979 these broad details were confirmed, and a more detailed picture dramatically emerged. The dark surface of Callisto (below left) is seen to be densely covered with impact craters. Above Callisto is Ganymede, which has a few impact craters visible as white patches with radiating rays, but is mainly covered with a curious marbled surface of rounded darker patches crossed by lighter striations.

Above Ganymede is Europa, and at this resolution one or two possible impact craters are visible, but the surface is still dominated by the broad dark areas of the equatorial regions and the lighter poles. Superimposed over these are a series of peculiar fine dark lines, up to thousands of kilometres long and remarkably straight over this distance. The origin of these remains a mystery, as this picture is the highest resolution yet available for Europa; before this book is published, however, Voyager 2 will, it is hoped, have photographed the satellite at much closer range.

Finally, at top left, is Io, the nearest of these satellites to Jupiter. On its surface no impact craters are visible, although there are many dark red spots, sometimes surrounded by bright haloes.

Satellites of Jupiter: Callisto and Ganymede

The differing numbers of impact craters on each of Jupiter's satellites give clues to the ages of their surfaces. The heavily cratered globe of Callisto suggests that it preserves a surface from the early history of the Solar System. Similar heavily cratered regions of the Moon yield dates of 4,000 million years or earlier. The most unusual feature on Callisto is the vast white area at lower left, 600km across and surrounded by several concentric rings, the largest of which is 2,600km across. It is reminiscent of the Orientale basin on the Moon and the Caloris basin on Mercury, but neither of the latter have so many concentric rings, and there are no radial striations or secondary craters visible in the basin opposite. The density of Callisto suggests that it is composed of both ice and rock, and it may be that such a mixture gives rise to the unique characteristics of this large basin.

Ganymede's surface is younger than Callisto's, and the number of impact craters seems similar to that of the lunar *maria*, implying an age for the present surface of about 3,000 to 3,900 million years. In the two pictures on the right (the top one covers 1,000km × 600km), more details of the curious striated patterns of Ganymede can be seen. These striations are quite unlike anything found on the Moon, Mars, Mercury or the Earth, so their origin is something of a puzzle at present. Close-ups (top left shows a region 350km × 670km) do not help, merely showing that the grooved patterns cross one another at fairly random angles, discounting any kind of flowage under gravity. It has been suggested that they might be the result of 'ice tectonics', as ice might produce significantly different patterns than does tectonic activity in rock, which on Mars has produced the spectacular features shown on pages 141 to 153. The grooved areas are noticeably lighter than the areas which have no grooves, perhaps because the ice is cleaner. It has also been suggested that the broad features of Ganymede's surface are what might be expected from continental drift driven by convection in the interior. If this idea is confirmed, Ganymede will be the only body so far explored, apart from the Earth, to have had a system of plate tectonics. Such a system must have ceased to be active long ago, probably more than 3,000 million years ago, as the number of superimposed impact craters implies.

As the Voyager 1 craft approached Europa and Io, it was expected that impact craters would become visible, as they are a dominant feature on all other terrestrial bodies whose surfaces have so far been photographed. Europa may have a few impact craters, but it was a great surprise that on Io, although a number of peculiar dark patches were seen (top left overleaf), there were certainly no impact craters. This means that there is a progression in surface age of the four satellites as one approaches Jupiter: Callisto is by far the oldest, Ganymede rather younger, Europa younger still and Io apparently extremely young.

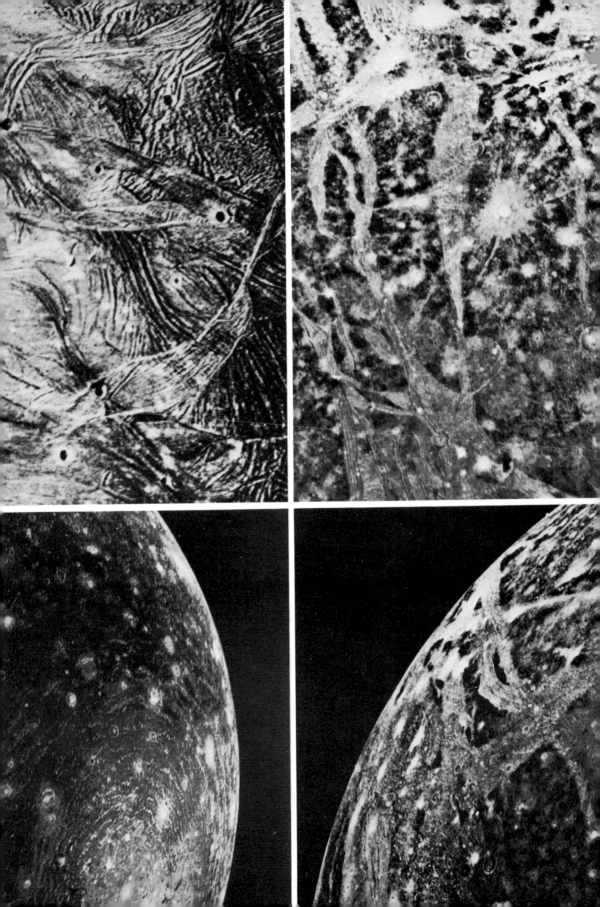

Satellites of Jupiter: Io

The greatest surprise was still to come. The limb of Io was regularly scanned, and the results used in a computer programme to help fix the spacecraft's position. On one of these routine scans, the computer was unable to fit a circle to the limb readings, and repeatedly refused the data. The appropriate data was made into a picture, and on the limb of Io was found a feature similar to that shown at top right. For thirty hours the Voyager team debated what the feature was, and finally made the stunning announcement that Io has erupting volcanoes. Six more were discovered, spaced around the planet near the equator. The middle right picture shows two of them: one causing the bright patch after sunset on the night hemisphere, the other causing the great dome of ejected matter 260km high on the lower right limb. These curious umbrellas are formed because Io has virtually no atmosphere and therefore no wind. The material thrown out from the volcano follows near-perfect parabolae, which together form the wide fountain seen best at top right. On Earth the finer material would be carried away downwind, and deposited widely over the surface. As closer pictures were taken of Io, the origin of the dark patches became clear, as seen in the middle left picture. The ragged fingers of dark matter spreading out from them are clearly lava flows, and closer still (bottom left) the outline of a crater about 50km across can be seen, with apparent lava flows radiating from it. The non-circularity of the craters marks them out as volcanic in origin rather than impact, and there is a strong resemblance to volcanic calderas on Earth. The caldera of Mauna Loa, in Hawaii, is shown at bottom right for comparison (the picture shows an area measuring about 6km × 7km).

Active volcanoes anywhere else in the Solar System would have been exciting enough, but Io, scarcely bigger than the Moon, is one of the last places we would have expected to find them. Yet, strangely enough, they were predicted in a paper by Peale, Cassen and Reynolds published only a few days before they were discovered. It was calculated that Europa and Ganymede would perturb the orbit of Io, causing the huge tidal bulge raised by Jupiter to vary, the friction causing internal heat great enough to cause melting. This would provide a perpetual means of internal melting, allowing the volcanoes to keep erupting and renewing the surface material, explaining the youth of Io's surface as evidenced by the lack of impact craters.

The discovery of active volcanoes on Io is probably the single most spectacular discovery in the whole of the exploration to date of the Solar System, and the fact that it was made only a few days before these pages were written is a reminder that planetary geology is a youthful and rapidly developing science; much remains to be explored and discovered. More detailed and extended views of Jupiter's satellites are to come in the Galileo mission; and we still have the surfaces of Venus, Titan, Triton and Pluto to provide us with more surprises and excitement in the years to come.

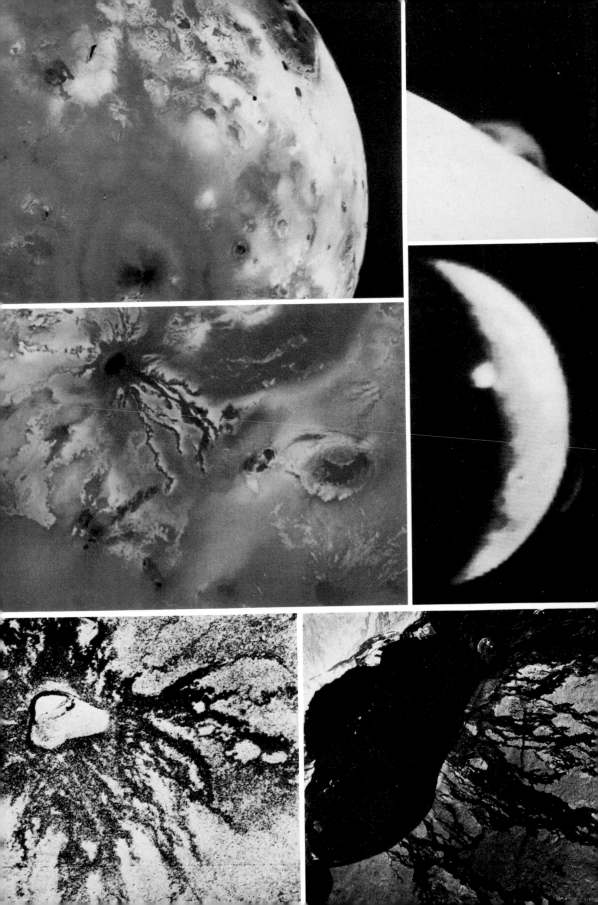

Acknowledgements

This book is almost entirely illustrated with photographs taken by the United States National Aeronautics and Space Administration. The illustration on p. 17 was taken from the *Consolidated Lunar Atlas*; the illustration on p. 197 (upper) was kindly provided by M. Malin and that on p. 197 (lower) by C. P. Florensky. The authors would also like to thank Pat Molloy and Joan Willsher for typing the manuscript, and David Rooks and John Barrett for preparing all the photographs.

Index